Anonymous

Report of the American Humane Association on Vivisection

in America

Adopted at Minneapolis, Minn., September 26, 1895. Vol. 3

Anonymous

Report of the American Humane Association on Vivisection in America
Adopted at Minneapolis, Minn., September 26, 1895. Vol. 3

ISBN/EAN: 9783337254681

Printed in Europe, USA, Canada, Australia, Japan

Cover: Foto ©berggeist007 / pixelio.de

More available books at **www.hansebooks.com**

REPORT

OF THE

American Humane Association

ON

VIVISECTION IN AMERICA

ADOPTED

AT MINNEAPOLIS, MINN.

SEPTEMBER 26, 1895.

———•———

CHICAGO, ILL.
THE AMERICAN HUMANE ASSOCIATION.
560 WABASH AVENUE.
1896.

University Press:
John Wilson and Son, Cambridge, U.S.A.

REPORT

OF

THE AMERICAN HUMANE ASSOCIATION

ON

VIVISECTION IN AMERICA.

MINNEAPOLIS, SEPT. 26, 1895.

———————

THE Special Committee appointed at the last annual meeting of the AMERICAN HUMANE ASSOCIATION, for the purpose of obtaining a census of the opinions regarding Vivisection which generally prevail, have now completed their task, and beg leave to submit the following report: —

The purpose and scope of their investigations seem to have been somewhat misunderstood. The American Humane Association did not wish to obtain a preponderance of signatures, either for or against the practice of vivisection. What seemed desirable was an investigation regarding the extent to which different views, particularly regarding painful experimentation, are now entertained by those more influential classes of society whose judgments exert the greater potency in the formation of public opinion. To do this, it became necessary to formulate precise statements of slightly diverging beliefs, in such form as that they should at any rate touch this one question of restriction or non-restriction of experiments in pain. It needs to be remembered that the word "vivisection," when used as a synonym for scientific experiments upon animals, may cover operations not more painful than a pin-prick; or, on the other hand, experiences as excruciating as the imagination can conceive. To ask simply whether one approves of vivisection or condemns it, would be meaningless, unless the definitions given were precise.

Three leading views regarding the practice of vivisection may be easily recognized: —

1. Its total condemnation because of tendency to cruelty.
2. Its restriction within certain limits.
3. Its approval without any other restraint than the will of the experimenter himself.

The second view, however, is capable of considerable subdivision. One person may favor vivisection provided it be limited to absolutely painless experiments. Another would not condemn it if the pain were slight, and the possible utility to humanity very great.

Four statements of opinion were carefully formulated, and to these a fifth was finally added, which was sent, however, as hereafter noted, to but very few others than members of the medical profession. But even these could by no means express all the shades of differing opinion which seem to exist in regard to this subject. Some few persons indeed could find nothing in either statement to endorse or condemn, while others, reading with extreme and careful discrimination, indicated by erasures and changes in phraseology their variance of opinion. But that the statements, as drawn, do fairly represent existing opinions is evinced by the fact, that, without any erasures or changes, each one received the signatures of men of national reputation.

The choice of persons whose views on this question should be solicited was a matter of no little consideration. It was, of course, impossible to ask all whose judgment would be valuable; and such selection was necessary as at least should be fairly representative in character and weight. It may be of interest to know the leading rules which governed the final choice of names.

1. In the first place, the opinions of the Medical Profession were desired. For their benefit, real or supposed, the practice of vivisection is mostly carried on. If, then, all experiments upon living animals are of such great value that no restraint whatever need be observed, it might be safely assumed that medical men after years of practical experience in the treatment of disease would be certain to know it. But the number of men in the medical profession in this country alone is over one hundred thousand, and to ask the judgment of all these would be manifestly impracticable. Then, too, the opinions of young men fresh from the medical school could not compare in value with those of other physicians who for years have been engaged in the combat with disease. It was necessary somewhat to draw lines. The medical profession in the States of New York and Massachusetts may be assumed to be as intelligently acquainted with the whole subject as physicians anywhere in the United States, and it seemed to us exceedingly probable that the views of physicians in other sections of the country would not greatly differ in their proportions from the varying judgments regarding vivisection which should be elicited in these two great common-wealths. Every physician in these two States who had been at least fifteen years in the practice of his profession was invited to give the Association the benefit of his mature experience on the question. A

few prominent medical men in other States received like invitations, but these instances were rare. The physicians upon the Board of Instructors in several leading medical colleges in New York, Boston, Philadelphia, Syracuse, Chicago, and other cities were also asked for expressions of opinion. The jury in this case seems to us to be fairly representative of the medical profession, and the results obtained were so unexpected as to call for special remark. If we may judge from the replies received, the majority of physicians who have tested value by experience are *not* in favor of "unrestricted vivisection."

2. For the opinions of those engaged in Educational work, our circulars were sent, first, to the president of each college and university in the United States; and, next, to the principal members of the teaching faculty of the leading institutions of learning.

3. In respect to Clergymen, it was very difficult to determine from whom among them opinions should be asked; and, as in the case of physicians, it was deemed best to solicit views only among those of the various denominations whose years of service or whose prominent position in the Church lent to their words a greater weight. To professors in various theological schools, to clergymen in our larger cities, and especially to men upon whom for years of service or unusual activity in clerical work the degree of Doctor of Divinity has been conferred, our circulars have been principally addressed. Your committee regret the modest reticence which in so many instances probably prevented that mention of honorary degrees to which signers were entitled, and of which note was requested. In some cases, by careful research, these have been supplied; but there are doubtless many to which such titles belong, and from which they are absent in the lists which follow.

4. From Literary Men, Editors, and Authors, opinions were sought, when addresses could be obtained, wherever achievement and worthy popularity lent weight to the expression of their views. Besides all these classes, there were a few others in other walks of life for whose judgment the world has regard, and whose opinions have been sought. From foreign countries also some expression of sentiment was desired; but these names are reckoned separately in the lists which follow.

Your committee believe that opinions thus obtained are of special value. We have not attempted a mere "counting of heads." We believe that the views thus collected do represent, in their proportions, the various shades of opinion which prevail among the educated classes of the country in regard to the question of unrestricted vivisection.

It is to be regretted that in some instances the purpose of this inquiry should have been so misunderstood, and that those who might have aided us by some expression of opinion preferred rather the expression of criticism or abuse.   One of the most prominent biologists of America suggested, with ghastly and peculiar humor, "that the American Humane Association do science the good service of suppressing anti-vivisectionists, — by persuasion, if possible, but *by vivisection*, if the seat of disease is too deep for more humane relief."   Another eminent scientific man belonging to Syracuse University went so far as to write an anonymous letter to your committee, suggesting that, before touching the question of vivisection, the American Humane Association should bring law to interfere with such cruelties as "poisoning roaches and bed-bugs, chasing butterflies, breaking down spiders' webs, stealing honey from bees," etc., — actions which he would apparently have regarded as even more reprehensible than the torments inflicted by Magendie, Mantegazza, and their imitators, upon man's nearest companion and friend.   Detection of the writer was not at all difficult, but further publicity than this would be too severe a punishment for what perhaps was a thoughtless act.

Numerous letters were received either accompanying signatures to one of the statements or in explanation of disagreement with their phraseology.   Your committee exceedingly regret that limitations of space prevent anything but the briefest quotation from the majority of these communications.   Some of these valued letters may perhaps hereafter be printed in other form.   For all suggestions and advice the committee tender thanks.   Even the few ungracious and critical epistles received are not without their value, if only as indications of proclivities and tendencies of which note must be taken.

What are the results of our investigations?   May we assume that in the more educated classes of the United States a majority of persons approve of the dissection of living animals to any extent and for any purpose a physiologist may desire?

No.   Of the total number of American opinions thus obtained there were : —

	Number.	Per Cent.
In favor of unlimited vivisection	281	13.4
Against unlimited vivisection	1,753	84.1
Evasive or obscure	52	2.5
Total	2,086	100.0

But vivisection is a practice with which physicians are principally concerned.   How do they stand?   Youth and enthusiasm would perhaps demand liberty of action in every way; it is only when

men have years of practical experience that questions of use or uselessness really arise. What is the judgment of medical men whose
experiences with the problems of disease and pain have taught them
where best to look for useful knowledge ? May we infer that as a
rule the medical profession demands the privilege of vivisection
without limitations or restrictions ?

No. So far as given, their vote is as follows : —

	Total No.	Per Cent.
In favor of unlimited vivisection	243	19.6
Against unlimited vivisection	968	78.1
Evasive or obscure	28	2.3
Total	1,239	100.0

We regard this result of our inquiries as of unusual value. For
many years it has been taken for granted that on one side of this
question of free vivisection were the only true guardians of scientific
research, the united ranks of physicians and surgeons, urging the
value of all experiments the physiologist desires to make; while
on the other side were to be found only ignorance and sentimen-
tality united in opposing what they did not comprehend and could
not value. We see that this is entirely an erroneous conception
of facts. It is only the minority of experienced physicians that
demand for physiological research unlimited opportunity independent of utility, and many of these are physiologists rather than
general practitioners. The great majority of medical men ask,
indeed, that experimentation shall not be wholly abolished; but.they
ask also *that it shall not be abused.* They condemn the cruel freedom
which exists on the continent of Europe, and, as a rule, the repetition
of painful experiments for the teaching of well-known facts.

We cannot refrain from quoting here the opinion of one of the
most distinguished physicians in this country, — Dr. Theophilus
Parvin, LL.D., late president of the American Academy of Medicine,
and professor in Jefferson Medical College, — as an illustration of
what we believe to be the position of the majority of the medical
profession in this country who have reached middle age : —

"It is wise for physicians interested in vivisection to recognize
that there is on the part of prominent women and men in the laity
a strong sentiment of antagonism to experiment on animals; and
therefore they should avoid all such work not promising certain
benefit to man, and anæsthetics ought always to be employed. I
sometimes fear that the anæsthesia is frequently nominal rather
than real, else why so many and ingenious contrivances for con-
fining the animal during operations, — contrivances that are not

made use of in surgical operations upon human beings, their immobility being secured by profound anæsthesia.

"Should the law restrict the performance of vivisection? I think it ought, chiefly as an expression of public sentiment, and for moral effect. . . . That restriction ought to forbid all experiments made without worthy objects, and in every case, so far as possible, the animal during and subsequent to the operation must be preserved from pain. . . .

"Vivisection is in more danger from ignorant, rash, and reckless experimenters than from those directly hostile to it. I cannot think that vivisections done for teaching purposes, simply showing what has been proved time and again upon hundreds and thousands of victims, are justifiable unless anæsthesia is employed, *to not merely mitigate, but to completely abolish suffering of the animals. Otherwise the influence of such experiments is injurious both to the operator and to the witnesses of the operation.*"

Let us analyze yet more closely the medical opinions received from Massachusetts and New York. The replies received from physicians in these States may be tabulated as follows:—

	Total No.	Per Cent.
For vivisection without restriction	220	19.1
For vivisection when restricted by utility	513	44.6
For vivisection when without pain	186	16.2
For the total prohibition of vivisection	207	18.
Obscure or evasive	24	2.1
Total	1,150	100.0

We see no reason to doubt that, with slight variations only, these proportions represent the sentiments prevailing in the medical profession throughout the country. If we ask the judgment of other influential classes in the community, we find the same tendency, even more pronounced. In the following tables we have indicated a number of opinions from clergymen, educators, presidents of universities and colleges, and those engaged in college work, etc:—

	Clergy-men.	Educa-tors.	Authors, Editors, etc.	Per Cent.
For vivisection without restriction . . .	0	34	4	4.7
For vivisection when restricted by utility .	189	84	63	41.2
For vivisection when without pain . . .	116	49	26	23.5
For the total prohibition of vivisection . .	144	52	30	27.7
Obscure or evasive	6	16	2	2 9
Total	455	235	125	100 0

To your committee, therefore, it seems certain that the majority of those who represent enlightened public opinion in this country are not favorable to unlimited vivisection. They do not indeed demand abolition of all experimental research whatever; nor are they agreed as to the scope of restrictions, the value of legal regulation, the definition of "utility," or the methods by which even desirable restraints may be practically enforced. There is also somewhat of scepticism as to the extent to which vivisection has been at all abused, and many who would quickly condemn the evil are far from being convinced that such evil exists. But one fact stands clearly out, that a majority of the class to whom we have appealed agree *that wherever vivisection approaches cruelty and uselessness, it should be prohibited and condemned by law.*

It is true, however, that this investigation has made evident the existence of a minority contrary sentiment, of which due note must be taken. There are physicians and surgeons whose self-sacrificing devotion to humanity is beyond the possibility of doubt, who have signed for unlimited vivisection without criticism or restraint. There are presidents of universities whose zeal for science has led them to declare that with vivisection "morality has nothing to do." There are clergymen of high repute who would condemn all cruelty in the abstract, but whose faith "in the substance of things unseen" would make of the physiological laboratory a temple for Humanity where all sacrifice is painless and its object the universal good.

There are large numbers of persons who suggest that all criticism of scientific methods is premature, or out of place in the presence of such other forms of wide-spread cruelty as accompany butchery and so-called sport. There are others who deprecate the infliction of pain or disuse of anæsthetics, unless it happens to be "necessary for the *success* of the experiment," — a qualification which at once nullifies the very restriction implied. In many colleges and institutions of learning we note a prevalent disposition to trust a vivisector entirely, "unless there seems reason for distrust," — a contingency which is not likely to happen, where trust is absolute and faith serene. But above everything else exists doubt or ignorance of what vivisection really is. A method of scientific research is shrouded in mystery, and is almost unknown and unknowable save to a few.

What action may be suggested to the American Humane Association as a result of your committee's inquiry?

The following conclusions seem to them not only to be justified by evidence, but in accord with the interests of both science and humanity: —

I. Vivisection is not merely a method of scientific teaching or investigation, but a practice which is justly subject to ethical restraints.

II. We believe this practice has been abused. We are compelled to admit that President Parvin was right in declaring before the American Academy of Medicine that there are some American vivisectors "who seem, seeking useless knowledge, to be blind to writhing agony and deaf to the cry of pain, and to have been guilty of the most damnable cruelties."

III. We believe in the potency of legislation to lessen these abuses. Again we agree with President Parvin that "law should restrict the performance of vivisection;" that "vivisections done for teaching purposes, simply showing what has been proved time and again upon hundreds and thousands of victims, are not justifiable unless anæsthesia is employed, to not merely mitigate, but to completely abolish the suffering of the animals." We are glad to find ourselves in perfect accord with a scientific authority who declares that "the influence of such experiments without anæsthesia *is injurious both to the operator and to the witnesses of the operation.*"

IV. But far transcending in importance the enactments of any restrictive legislation is the wide dissemination of absolute, accurate knowledge of vivisection as it is to-day carried on in the seclusion of American laboratories. It is alleged that the infliction of pain is of rare occurrence; that abuses are unknown, and that nothing whatever is done which in any way needs to be concealed from the public eye. But these statements are controverted. How then shall truth be reached and be made evident? It seems to us that knowledge, if attainable, must rest here, as Science declares it must rest elsewhere, not on the statements of interested parties. but on that firm basis she herself demands, — the accurate observation of facts by impartial witnesses. Yet, how is such evidence possible so long as the doors of laboratories are locked and barred to the public, and even to physicians and surgeons who are not personally known? In the interests of scientific truth we suggest therefore that an experiment be made. We recommend that the American Humane Association during the coming year ask of each physiological laboratory in this country, whether it will accord permission to the President of the local Humane Society or to his authorized representative to be present during any experiments upon animals that may take place, simply as a silent observer, and entirely without privilege of suggestion, criticism, or unsolicited remark. If there be no occasion for mystery or seclusion, we believe this privilege will be accorded, if only in the interests of truth and for the dissipation of error. If refused, there can be but one inference and but one remaining appeal.

Respectfully submitted.

(Signed)   TITUS MUNSON COAN, M. D.
ALBERT LEFFINGWELL, M. D.
MATTHEW WOODS, M. D.

In accordance with the above recommendation the American Humane Association adopted the following resolution: —

"*Resolved*, That the American Humane Association, having considered the report of its Special Committee on Vivisection, believes that it should ask for yet more light regarding the aims and methods of experimentation upon living animals. Laying aside for the present all other evidence, the Association seeks now for positive knowledge by that method which Science idealizes and demands, — based, not upon authority, but upon the accumulated experience of a large number of independent observers in all parts of the country.

"The Association therefore instructs its President and Secretary, as soon as may be, to ask the authorities having in charge every known physiological laboratory in the United States, whether or not they will accord permission to the President of the local Society, having for its object the prevention of cruelty, or to his authorized representative for the time being, the right of admission to the laboratory during any experiments of any kind upon living animals, simply as a student and observer, and *entirely without the right of suggestion, criticism, or unsolicited remarks of any kind.*"

I. ABSOLUTE PROHIBITION OF VIVISECTION.

ALL experimentation upon living animals we consider unnecessary, unjustifiable, and morally wrong. Some of the highest medical authorities have asserted that Vivisection does not benefit mankind; that owing largely to differences between the structure of men and animals, the results of operations and the test of drugs upon the latter are wholly misleading, and that the practice has accomplished nothing of real value in the treatment of disease. The greatest physiologist of our century, Sir Charles Bell, declared that such "experiments have never been the means of discovery;" and that "the opening of living animals has done more to perpetuate error than to enforce the just views taken from anatomy and natural science." Mr. Lawson Tait, one of the most eminent of living surgeons, claims that but for the fallacies of Vivisection, the art of healing would be to-day "at least a century in advance of its present position;" and Dr. Bell Taylor, one of the leading oculist-surgeons of Great Britain, affirms that "no good ever came from the practice, and no good ever will." With these scientific authorities we are in perfect accord.

But whether any useful knowledge can be thus acquired or not is beside the question. Even if utility could be proved, man has no moral right to attempt to benefit himself at the cost of injury, pain, or disease to the lower animals. *The injury which the practice of Vivisection causes to the moral sense of the individual and to humanity far outweighs any possible benefit that could be derived from it.* Dr. Henry J. Bigelow, Professor in the Medical School of Harvard University, declared that "*Vivisection deadens the humanity of the students.*" Nothing which thus lowers morality can be a necessity to progress.

We hold the infliction of torture to be a moral offence; and believe experience has demonstrated that Vivisection cannot be sanctioned in any form without opening a door to that offence. We assert that no legal protection from the utmost extremity of torment can ever be given to an animal when once it is laid on the Vivisection table in the laboratory; and that no line can be drawn between experiments that are painless and those involving the utmost torture. The claim that Vivisection has been rendered generally painless by the use of anæsthetics is wholly misleading; for physiologists themselves admit that there are no less than thirteen classes of experiments which cannot be satisfactorily performed on anæsthetised animals.

To allow one kind of Vivisection because "painless" and to condemn another because "painful" is thus utterly impracticable. No distinctions could be drawn that the enthusiastic experimenter would regard, and no legal restrictions are possible that would be conscientiously observed. Painless or painful, useless or useful, however severe or however slight, Vivisection is therefore a practice so linked with cruelty, and so pernicious in tendency, that any reform is impossible, and it should be absolutely prohibited by law for any purpose.

Prof. JAMES E. GARRETSON, M. D., Senior Professor of Surgery, Medico-Chirurgical College, Philadelphia : —

"I am without words to express my horror of vivisection, though I have been a teacher of anatomy and surgery for thirty years. It serves no purpose that is not better served after other manners."

FORBES WINSLOW, D. C. L. Oxon., M. R. C. P., London, Editor Journal Psychological Medicine; Physician to the British Hospital for Mental Diseases, Physician to North London Hospital for Consumption, etc. : —

"In my opinion, vivisection has opened up no new views for the treatment and cure of diseases. It is most unjustifiable and cruel, and in no way advances medical science. I do not believe in many of the so-called experiments. made by these 'faddists,' especially those relating to brain operations on monkeys and the consequent theory of cerebral localization. I have probably more experience than many of these experimenters who have given their opinions to the world as based on what they have done; and I beg leave to express my utter disbelief in the usefulness of such experiments, and to discredit their being followed by any good results to mankind or to science in general."

Rt. Hon. A. J. MUNDELLA, F. R. S., London.

ALFRED RUSSELL WALLACE, F. R. S., D. C. L. Oxon., LL. D., Author and Naturalist : —

"I would have signed II. or III. but that I consider effectual official regulation impossible, and the increase of official inspectors altogether impolitic."
(After the first word he inserts "painful or injurious.")

EDWARD BERDOE L. R. C. P. E., M. R., C. E., Physician and Surgeon, London.

Hon. JUSTIN McCARTHY, M. P., Author, London.

Prof. WILLIAM J. MORTON, M. D., Professor of Nervous and Mental Diseases at the New York Post-Graduate Medical School and Hospital, New York City : —

"I only wish I could state the above sentiments stronger. If mankind suffers from disease it is its own fault, to be cured by rectification of the causes which lead to it; and it is subversive of the high and moral order of the progress of humanity to inflict pain or death upon other living animals to abolish or minimize disease or suffering due to mankind's own faults. In the end, the retribution to the race which does this will equal and offset the advantages temporarily gained. One crime or fault does not excuse or justify another."

(To Dr. Morton's father, Dr. W. T. G. Morton, the world owes one of the greatest blessings of this or any other age, — the comparative conquest of pain by the inhalation of ether.)

B. F. SHERMAN, M. D., Ex-President of the New York State Medical Society, Ogdensburg, N. Y.: —

"If it could be restricted to utility and without pain, it would be all right; but if permitted at all, it will be abused."

EDWIN A. W. HARLOW, A. M., M. D. (Harvard), Wollaston, Mass.:

"The late Dr. Henry J. Bigelow, in a lecture, which I heard, before the Harvard College Medical School, condemned the practice of some of the students in Paris in their vivisections on horses, without anæsthetics, as 'infernal inhumanity.' Vivisections in all Medical Schools should be abolished."

E. H. HAWKS, M. D., Lynn, Mass.: —

"I believe that vivisection blunts the moral sense to such a degree as to become a strong force in the production of criminals."

J. D. BUCK, M. D., Professor of Nervous Diseases and the Principles of Therapeutics, and Dean of Pulte Medical College, Cincinnati, Ohio.

ELMORE PALMER, M. D., President (1890) of the Western New York Medical Society, Buffalo, N. Y.

WILLIAM INGALLS, M. D., Boston, Mass.: —

"*Absolute prohibition;* for unless a law can be made which no one can get away from, vivisection will obtain just as it does now."

ALLAN MOTT-RING, M. D., Arlington Heights, Mass.: —

"Vivisection is an unmanly crime."

IRA CLARK GUPTILL, M. D., M. S., Northborough, Mass.: —

"No legal restrictions would be conscientiously observed, and therefore I strike for absolute prohibition by law."

ALEX. S. McCLEAN, M. D., Springfield, Mass. : —

"Have been in practice forty-eight years, and have never been influenced or governed by anything I have seen or read in the line of vivisection."

LORENZO W. COLE, M. D., Springfield, Mass. : —

"I consider it barbarous to torture anything capable of feeling pain, to demonstrate facts which have been proven thousands of times."

Rt. Rev. JOHN SCARBOROUGH, D. D., Bishop of New Jersey.
Rt. Rev. JOHN WILLIAMS, D. D., LL. D., Bishop of Connecticut.

Rt. Rev. Hugh Miller Thompson, D. D., Bishop of Mississippi.

Rt. Rev. J. H. D. Wingfield, D. D., Bishop of N. California.

The Very Rev. E. A. Hoffman, D. D., D. C. L., Dean of the General Theological Seminary, New York.

Rev. Dr. Lyman Whitney Allen, Newark, N. J. : —

"If vivisection could be absolutely without pain, I should be willing to advocate it under certain restrictions. But who shall be the judge ? As it is at present conducted, it is immoral and wicked. The only safe and Christian standpoint is absolute prohibition. Science has its limits, as have other branches of investigation. Valuable and important as is biological study, life and animal individuality and consciousness have their God-given rights."

Rev. Dr. G. II. De Bevoise, Keene, N. H. : —

"Although never having occasion to protest against an act of this kind because of knowledge of its occurrence, I am most decidedly in favor of the absolute prohibition of vivisection, if this may be accomplished. The end in view by no means justifies the means, and very often the enthusiasm of scientific investigation overrides all humane considerations, and is willing that there may be any amount of suffering if only ambition may accomplish its end. Vivisection is more than cruel. It is wicked cruelty, concentrated."

Rev. Dr. Amory H. Bradford, Associate Editor of "The Outlook," New York : —

"I incline to a middle ground between I. and II., but prefer to sign the first."

President Henry Morton, Ph. D., Stevens Institute of Technology, Hoboken, N. J.

President David C. John, D. D., Clark University, Ga.

President G. W. Holland, Ph. D., Newbury, S. C.

President Lawr. Larsen, Norwegian Luther College, Iowa.

(Omits first paragraph, the reference to Dr. Bigelow, and the last sentence of the third paragraph.)

President McK. H. Chamberlin, LL. B., McKendree College, Illinois.

Samuel S. Garst, M. D., Ph. D., Ashland University, Ohio.

President L. L. Hobbs, A. M., Guilford College, N. C.

President Jesse Johnson, A. M., Muskingum College, Ohio : —

"No set of men may limit the application of the moral law."

President John Van Ness Standish, LL. D., Lombard University, Ill.

President John Braden, Central Tenn. College, Nashville, Tenn.

President William F. Shedd, D. D., Acting President of Little Rock University, Little Rock, Ark.

President H. J. Kiekhoefer, A. M., Northwestern College, Ill.

President E. BENJAMIN BIERMAN, Ph. D., Lebanon Valley College, Penn.

President C. W. CARTER, D. D., Centenary College, La.

President W. F. MELTON, Ph. D., Florida Conference College, Fla.:
"Vivisection without restrictions is devilish."

President W. P. JOHNSTON, D. D., Geneva College, Penn.

President TAMERLANE P. MARSH, Mt. Union College, Albion, Ohio: —
"My judgment is based on the accuracy and authority of the scientists quoted. If inaccurate, I should sign No. III."

Prof. J. P. WIDNEY, President of the University of Southern California, Los Angeles, Cal.

Prof. FRANCIS J. WAGNER, A. M., D. D., President of Morgan College, Baltimore, Md.

Rev. J. C. CLAPP, D. D., President of Catawba College, Newton, N. C.

Rev. JAMES W. STRONG, President of Carleton College, Northfield, Minn.

Rev. WM. HENSLEE, President of Pierce Christian College, College City, Cal.

Rev. MERLE A. BREED, President of Benzonia College, Benzonia, Mich.

President ISAAC N. RENDALL, Lincoln University, Lincoln, Pa.

President GEORGE HINDLEY, Ridgeville College, Ridgeville, Ind.:
"Absolute prohibition of vivisection until *man is absolutely regenerated.*"

President JAS. T. COOTE, A. M., Washington College, Tenn.

President J. P. GREENE, D. D., LL. D., William Jewell College, Liberty, Mo.

President CHARLES A. BLANCHARD, Wheaton College, Wheaton, Ill.

WILLIAM H. PAYNE, LL. D., Chancellor of the University of Nashville, Tenn.

Prof. HIRAM CORSON, LL. D., Professor of English Literature, Cornell University, Ithaca, N. Y.

Prof. CHARLES MELLEN TYLER, A. M., D. D., Professor of Christian Ethics, Cornell University, Ithaca, N. Y.

Prof. W. S. TYLER, D. D., LL. D., Professor of Greek, Amherst College, Mass.

Prof. HARRY T. PECK, A. M., Ph. D., Professor of Latin Language and Literature, Columbia College, N. Y.

Prof. G. C. WHEELER, B. S., Ph. D., Chair of Chemistry, Cornell University, Ithaca, N. Y.

Prof. S. G. WILLIAMS, A. B., Ph. D., Professor of the Art of Teaching, Cornell University, Ithaca, N. Y.

Prof. A. M. Wheeler, M. A., Professor of History, Yale University, New Haven, Conn.

Prof. George B. Stevens, Ph. D., D. D., Professor of New Testament Criticism, Yale University, New Haven, Conn.

Prof. William Cleaver Wilkinson, University of Chicago.

Prof. Edward Lee Greene, Catholic University of America, Washington, D. C.

Prof. Arvin S. Olin, University of Kansas.

Prof. Edward E. Hale, Jr., State University, Iowa : —

"I should much prefer to sign II. or III., if it were not that I see no practicable way of restricting vivisection. If it were possible to draw the line between painful and painless vivisection, between needful and useless vivisection, it would be almost impossible to have the line observed. If any way could be shown that would surely accomplish this end, I should sign III."

(Professor Hale erases all statements of fact which he has not means of verifying.)

Prof. W. H. Savage, El Paso, Texas : —

"Under limitations it might do to delegate to a very few eminent men the right to make some experiments; but such experiments should not be made before a body of students. Even to advance science, we have no right to torture animals."

Prof. James Otis Lincoln, San Mateo, Cal.

Prof. William F. Phelps, M. A., St. Paul, Minn.

Charles W. Stone, A. M. (Harvard), Boston.

Rev. Dr. Hiram C. Hayden, Vice-President of the Western Reserve University, Cleveland, Ohio : —

"Signed on strength of the testimony of the medical experts, quoted above."

Prof. Nathan Abbott, LL. B., Professor of Law, Leland Stanford Jr. University, Palo Alto, Cal.

M. L. Holbrook, M. D. (Prof. Hygiene in New York Med. College, Editor "Journal of Hygiene,"), New York.

Julius F. Krug, M. D., Buffalo, N. Y.

Horace C. Taylor, M. D., Brockton, N. Y.

Isaac D. Meacham, M. D., Binghamton, N. Y.

Wm. P. Roberts, M. D., Minneapolis, Minn.

Herbert Beals, M. D., Buffalo, N. Y.

D. A. Dean, M. D., Buffalo, N. Y.

E. A. Burnside, M. D., Buffalo, N. Y.

O. M. Frisbie, M. D., Bainbridge, N. Y.

G. H. R. Bennet, M. D., Brooklyn, N. Y.

P. Pryne, M. D., Herkimer, N. Y.: —

"I am opposed to vivisection under all circumstances, without reservation."

Jesse Myer, M. D., Kingston, N. Y.: —

"The sentiments of the statement for 'Vivisection without restrictions' are appropriate to the Dark Ages. The barbarism of the Turk would blush to hear them."

Howard Wetmore, M. D., New York City.

Peter B. Wyckoff, M. D., New York City.

Thos. M. Dillingham, M. D., New York City.

Follansbee G. Welch, M. D., New York City.

Lewis Hallock, M. D., New York City.

Luigi Galvani Doane, M. D., New York City.

Jas. R. Bird, A. M., M. D., Brooklyn, N. Y.

Christian Enrich, M. D., New York City.

Addison C. Fletcher, M. D., New York City.

2

Jas. H. Patton, A. M., M. D., Brooklyn, N. Y.
William P. Morrissy, M. D., Brooklyn, N. Y.
B. Fincke, M. D., Brooklyn, N. Y.
Jas. E. Russell, M. D., Brooklyn, N. Y.
Jas. Murphy, M. D., Sherman, N. Y.
Jas. P. Hawes, M. D., North Hector, Schuyler Co., N. Y.
Hiram C. Driggs, M. D., New York City.
Wm. G. Hartley, M. D., New York City.
S. B. Childs, M. D., Brooklyn, N. Y.
John Cooper, M. D., M. R. C. S., Brooklyn, New York.
Geo. W. Newman, M. D., Brooklyn, N. Y.
B. B. Roberts, M. D., Buffalo, N. Y.
John F. Wage, M. D., Buffalo, N. Y.
F. W. Stilwell, M. D., Rochester, N. Y.
John H. Eden, M. D. (Yale, 1874), Fordham Heights, N. Y.
Thos. B. Heimstreet, M. D., Troy, N. Y.
C. E. Fraser, M. D., Rome, N. Y.
John Reid, M. D., Rochester, N. Y.
A. R. Green, M. D., Troy, N. Y.
Herman Beyer, M. D., Stapleton, N. Y.
W. W. Archer, M. D., Clifton Springs, N. Y.
Wilson T. Bassett, M. D., Cooperstown, N. Y.
B. C. Andrews, M. D., Dansville, N. Y.
David J. Mallery, M. D., Bristol Centre, N. Y.
B. D. Mosher, M. D., Granville, N. Y.
W. W. Whiting, M. D., Union, N. Y.
W. E. Gorton, M. D., Corning, N. Y.
Theodore C. Wallace, M. D., Cambridge, New York.
C. J. Farley, M. D., Sandy Hill, N. Y.
William C. Cooke, M. D., Moravia, N. Y.
Peter H. Hulst, M. D., Greenwich, N. Y.
Albert H. Crump, M. D., Williams Bridge, New York.
W. D. O. K. Strong, M. D., Fishkill-on-Hudson, N. Y.
C. V. H. Morris, M. D., Lodi, N. Y.
J. J. Allman, M. D., Union Springs, N. Y.
J. B. Hartwell, M. D., Woodsburgh, N. Y.
W. W. Budlong, M. D., Frankfort, N. Y.
Newton Cook, M. D., Sandy Creek, N. Y.
R. R. Thompson, M. D., Kingston, N. Y.
James H. Rogers, M. D., East Hampton, N.Y.:

" The vivisector is only less guilty than the man or woman who kills or maims an animal for amusement, or in obedience to the dictates of fashion."

Jos. Sidney Crane, M. D., Huntington, N. Y.
Charles T. Mitchel, M. D., Canandaigua, N.Y.
Horace Halbert, M. D., Canastota, N. Y.
Wm. G. Ware, M. D., Dedham, Mass.
C. K. Beldin, M. D., Jamaica, N. Y.
C. T. Greenleaf, M. D., Brewerton, N. Y.
A. J. Evans, M. D., Fredonia, N. Y.

Ezra McDougall, M. D. (Univ. Med. Coll. N. Y.), Oneonta, N. Y.: —
" I have never derived any benefit from the practice of vivisection."
Wm. M. Gwynn, M. D., Throopsville, N. Y.
J. Fenimore Niver, M. D., Cambridge, N. Y.
Ira D. Brown, M. D., Weedsport, N. Y.: —
" Every word in the above statement I know to be true. The practice of vivisection is inhuman, cruel, and brutalizing in its effects upon those who witness it, while no information useful to the human family is gained from it. In our medical colleges it is indulged in as a sport, a pastime, to the moral degradation of the students, making them unfit for the practice of the healing art."

Geo. A. Stuart, M. D. (Harvard), Oyster Bay, Queens Co., N. Y.
T. J. Peer, M. D., Ontario, N. Y.
T. J. Kilmer, M. D., Schoharie, N. Y.
Jas. S. Raymond, M. D., Ogdensburg, N. Y.
John Vedder, M. D., Saugerties, N. Y.
H. J. Nims, M. D., Manlius, N. Y.
Dr. L. Jeschinsky, M. D., Mt. Vernon, N. Y.
A. B. Carpenter, M. D., N. Greece, N. Y.
Chas. E. Smith, M. D., Whitesboro, N. Y.
Samuel S. Wallian, M. D., Helix, San Diego Co., Cal.
Sarah E. Wilder, M. D., Boston, Mass.
Alice Borle Campbell, M. D., Brooklyn, N. Y. (member of the consulting staff of the Memorial Hospital, Brooklyn, and member of the consulting staff of the New York Medical College and Hospital for Women).
Adele A. Gleason, M. D., Elmira, N. Y.
Sarah E. Bissell, M. D., New York City.: —
" Vivisection is useless, degrading, brutalizing."

Georgiana M. Crosby, M. D., New York City.
Eliza Ellinwood, M. D., Rome, N. Y.
Rachel T. Speakman, M. D., West Chester, Pa.
Jeannette B. Greene, M. D., New York City.
S. A. Skinner, M. D., Hoosac Falls, N. Y.
(Miss) S. S. Nivison, M. D., Dryden, N. Y.
Miles B. Jones, M. D., Camden, N. Y.
P. Rebekah Johnson, M. D., New York City.
Mary Slade, M. D., Castile, N. Y.
L. S. Sprague, M. D., Williamson, N. Y.
B. F. Underwood, M. D., Arlington, N. J.
D. S. Woodworth, M. D., Fitchburg, Mass.
I. Eaton Chase, A. M., M. D., Haverhill, Mass.
A. Houghton Kimball, M. D., Boston, Mass.
Francis M. Cragin, M. D., Norwood, Mass.
M. H. Williams, M. D., New York City.
John F. Miller, M. D., New York City.

W. H. Krause, M. D., New York City.
John D. Quackenbos, M. D., New York City.
James Neil, M. D., New York City (Examining Surgeon for the U. S. Pension Bureau from 1862 to 1867).
Franz Heuel, M. D., New York City.
M. A. B. Mount, M. D., New York City.
H. Clay Paddock, M. D., New York City.
Matthew Woods, M. D., Philadelphia, Pa.
T. Griswold Comstock, A. M , M. D., Ph.D., St. Louis, Mo.
W. A. Earle, M. D , Chairman of the Board of Health, West Boylston, Mass.
James P. Broderick, M. D., Jamaica Plain, Mass.
Daniel D. Slade, M. D., Chestnut Hill, Mass.
E. Herbert Noyes, M. D., West Gloucester, Mass.
G. Colburn Clement, M. D., Haverhill, Mass.
James P. Elliot, M. D., Boston, Mass.
Caroline E. Hastings, M. D., Boston, Mass.
Charles H. Shackford, M. D., Physician and Surgeon, Frost Hospital, Chelsea, Mass.
N. G. Macomber, M. D., Westport, Mass.
Francis J. Stevens, M. D., Boxford, Mass.
James D. Judge, M. D., Boston, Mass.
John B. Tyler, M. D., Billerica, Mass.
Alfred L. S. Morand, M. D., Jamaica Plain, Mass.
Drs. Sturtevant and Hallowell, Hyde Park, Mass.
James Warwick, M. D., Worcester, Mass.
J. B. Conklin, M. D., Albany, N. Y.
Sarah E. Sherman, M. D., Salem, Mass.
Julia B. Wood, M. D., Worcester, Mass.
Charlotte Evans Page, M. D., Lowell, Mass.
Harriet N. Watson, M. D., Albion, N. Y.
Wm. M. Moore, M. D., Provincetown, Mass.
J. G. Johnson, M. D., Wellfleet, Mass.
William R. Hayden, M. D., Bedford Springs, Mass.: —
   " Vivisection serves no good purpose; and I agree with Dr. Bigelow."
Henry R. Brissett, M. D., Lowell, Mass.
B. Hubbard, M. D., Plymouth, Mass.
Walter E. Harvey, M. D., Cambridgeport, Mass.
Walter S. Hall, M. D., Medford, Mass.
M. F. Delano, M. D., Late Surgeon of the U. S. Army, Sandwich, Mass.
L. V Gibbs, M. D., Worthington, Mass.
C. T. Stanley, M. D., Amesbury, Mass.
G. E. White, M. D., Sandwich, Mass.: —
   " The first statement more accurately meets my views than any other. There may be circumstances where it would be of great benefit."

John T. Harris, M. D., Boston, Mass.
J. B. Andrews, M. D., Lynn, Mass.
James B. Bell, M. D., Boston, Mass.
Homer S. Bell, M. D., Granby, Mass.
Byron R. Harmon, M. D., Woburn, Mass.
Charles Kingsbury, M. D., Boston, Mass.
George H. Payne, M. D., Boston, Mass. ·
Charles E. Nichols, M. D., Boston, Mass.
Marcus F. Bridgman, M. D., Boston, Mass.
E. J. Gooding, M. D., Boston, Mass.
Z. A. Spendley, M. D., Chenango Forks, N. Y.
Nathaniel M. Freeman, M. D., New York City.
George F. Oertel, M. D., New York City.
F. E. Martindale, M. D., Port Richmond, N. Y.
J. M. Turner, M. D., Lyons, N. Y.
Adelbert D. Head, M. D., Syracuse, N. Y.
Gilbert R. Traver, M. D., Perry, N. Y.
David Edward Collins, M. D., Medway, N. Y.
Thos. W. Bennett, M. D., Jeffersonville, N. Y.
Elizabeth H. Bates, M. D., Port Chester, N. Y.:
   " Operations can be done on dogs and other animals that owing to differences in organs cannot be done on human beings. Vivisection may show curious things, but at the expense of the inhumanity of man."

T. S. Quick, M. D., Carrollton, N. Y.
Chas. O. Hawkins, M. D., Brooklyn, N. Y.
James I. Marcley, M. D., Buffalo, N. Y.
Timothy Dwight Stow, M. D., Mexico, N. Y.
H. Hadley Smith, M. D., Hudson, N. Y.
A. B. Rice, M. D., Jamestown, N. Y.
J. F. Cleveland, M. D., Le Roy, N. Y.
A. C. Miller, M. D., Turin, N. Y.
John Murphy, M. D., New York City.
Fred C. Robinson, M. D., New York City.
Dr. Edward Frankell, M. D., Consulting Surgeon to the City Hospital, New York City.
J. W. Ferris, M. D., Mount Vernon, N. Y.
Alfred Large, M. D., Great Barrington, Mass.
Charles M. Brockway, M. D. (College Physician and Surgeon, New York), Worcester, Mass.
Thomas B. Shaw, LL. B., M. D. (Harvard), Lowell, Mass.
Edgar Leroy Draper, A. M., M. D. (Harvard), Holyoke, Mass.
John H. Keyser, Hartford, Conn.
M. M. Averill, M. D., Lynn, Mass.
Stephen Witt, M. D., North Dana, Mass.
S. Saltmarsh, M. D., Lexington, Mass.
G. W. Burdett, M. D., Clinton, Mass.
William P. Burge, M. D., Westfield, Mass.
C. H. Harriman, M. D., Whitinsville, Mass
A. C. Lane, M. D., West Medford, Mass.
Horace M. Nash, M. D., Lancaster, Mass.

Edward T. Tucker, M. D.; New Bedford, Mass.

Thomas A. Capen, M. D., Fall River, Mass.

Colby Lamb, M. D., Salem, Mass.

Benj. Benoit, M. D., Lowell, Mass.: —
"Have always looked upon vivisection as . cruel in the extreme."

Lorenzo Waite, A. M., M. D., Pittsfield, Mass.

William L. Johnson, M. D., Uxbridge, Mass.

Seth W. Kelley, A. M., M. D., Woburn, Mass.

C. W. Bowen, M. D., Westfield, Mass.

David W. Hodgkins, M. D., East Brookfield, Mass.

Rosto O. Wood, M. D., Worcester, Mass.

Charles F. Sherman, M. D., Haverhill, Mass.

George A. Coburn, A. M., M. D., Cambridge, Mass.

V. L. Owen, M. D., Springfield, Mass.

Isaac Thorndike Hunt, M. D., Boston, Mass.

Marshall L. Brown, M. D., Boston, Mass.

Rev. Dr. Robert R. Booth, New York City.

Rev. Dr. Arthur Brooks, New York City.

Rev. Dr. Chas. E. Robinson, Scranton, Pa.

Rev. Dr. J. Clement French, Newark, N. J.

Rev. Dr. James B. Gregg, Colorado Springs.

Rev. Dr. Burdett Hart, New Haven, Conn.

Rev. Dr. H. Nelson Hollifield, Newark, N. J.

Rev. Dr. John R. Davies, New York City.

Rev. Dr. John V. L. Reynolds, Meadville, Pa.

Rev. Dr. Levi Parsons, Mt. Morris, N. Y.

Rev. Dr. William C. Hopkins, Toledo, Ohio.

Rev. Dr. C. H. Rogers, Oklahoma City.

Rev. Dr. Alex Kent, Washington, D. C.

Rev. Dr. Francis E. Miller, Paterson, N. J.

Rev. Dr. John W. Brown, New York City.

Rev. Dr. Epher Whitaker, Southold, New York.

Rev. Dr. Robert Court, Lowell, Mass.: —
"My mind has hovered between I. and II.; but has finally resolved to I. as safest, most humane, and most in accordance with the mind of Him who 'made and loveth all.' "

Rev. Dr. Charles J. Jones, Stapleton, N. Y.

Rev. Dr. Dewitt M. Benham, Ph. D., Pittsburg, Pa.

Rev. Dr. D. Hall, Indiana, Pa.

Rev. Dr. James Henry Ecob, Albany, N. Y.

Rev. Dr. Rhys G. Jones, Utica, N. Y.

Rev. Dr. A. St. John Chambré, Arch-Deacon of Lowell, Mass.

Rev. Dr. James J. Burd, Utica, N. Y.

Rev. Dr. Joseph M. Clarke, Syracuse, N. Y.

Rev. Dr. Edmund B. Willson, Salem, Mass.

Rev. Dr. Edward T. Fairbanks, St. Johnsbury, Vermont.

Rev. John A. Bellows, Portland, Maine.

Rev. James Kay Applebee, Allston, Mass.

Rev. Otis A. Glazebrook, Elizabeth, N. J.

Rev. Geo. Augustine Thayer, Cincinnati, Ohio.

Rev. Frederick S. Sill, Cohoes, N. Y.

Rev. Ithamar W. Beard, Dover, N. H.: —
"I indorse this view, not only on account of the pain inflicted, but more especially because of the statement that 'vivisection deadens humanity.' I can conceive no benefit that can come to our race commensurate with this great harm."

Rev. Ernest Smith, M. A., Baltimore, Md.

Rev. James Holwell, Kioder, Owego, N. Y.

Rev. Robert Hudson, Syracuse, N. Y.

Rev. Stephen Peebles, Satauk, Colorado.

Rev. James H. Darlington, Brooklyn, N. Y.

Rev. Floyd W. Tompkins, Jr., Providence, R.I.

Rev. Dr. Joshua Young, Groton, Mass.

Rev. William Brunton, B. D. (Harvard), Whitman, Mass.

Rev. A. J. Chapin, Omaha, Neb.

Rev. E. M. Hickok, Sharon, Mass.: —
"No permanent good results can come to humanity from such violation of God's law of love."

Rev. John T. Rose, Cazenovia, New York.: '
"If it were within my power, I should absolutely prohibit vivisection, and visit the crime with legal and social penalties."

Rev. George G. Perrine, Guildford, N. Y.

Rev. T. Lewis Banister, New Hartford. N Y.:
"No human being has the right to inflict torture upon an inferior order of beings incapable of determining and expressing voluntary choice. If, however, any heroic *human* being in full possession of all his reasoning faculties, ascertained as such by such legal process as determines cases of insanity by due examination, who at the time of such examination, after being pronounced *sane*, declares his willingness to become a martyr to science, — *let him*."

Rabbi L. Weiss, Columbus, Ohio.

Rabbi Henry Cohen, Galveston, Texas.

Rabbi Max Wertheimer, Dayton, Ohio.

Rev. D. R. Hardison, Italy, Texas.

Rev. C. R. D. Crittenton, Chicago, Ill.

Rev. John C. Kimball, Hartford, Conn.
(Adds, "and by appeals to reason, sentiment, and moral principle.")

Rev. Stephen H. Camp, Brooklyn, N. Y.

Rev. James Huxtable, Boston, Mass.

Rev. James De Normandie, Boston, Mass.

Rev. Archdeacon Gilbert F. Williams, Washington, D. C.

Rev. Archdeacon J. D. Morrison, D. D., LL. D., Ogdensburg.

Rev. E. F. H. J. Massé, B. A., Oxon., Chicago.

Rev. Hobart Cooke, Plattsburg, N. Y.

Rev. George Clark Houghton, D. D., Hoboken.

Rev. Eaton W. Maxcy, Troy, N. Y.

Rev. Nathan F. Whiting. Cape Vincent, N. Y.

Rev. Dr. S. W. Miller, Saltsburg, Pa. : —

"For years I favored H. Could I believe it practicable, I might still favor it; but the seeming impossibility of making it so, the liability to abuse in the hands of unprincipled persons, and the fact that in many cases the use of anæsthetics defeats the purpose, compels me to adopt I."

Rev. F. N. Palmer, Pueblo, Colo.

Rev. Geo. W. Chalfant, D. D., Pittsburg, Pa.

Rev. Theodore F. Wright, Cambridge, Mass.

Rev. Joseph Dunn Burrell, Brooklyn, N. Y.

Rev. Randolph H. McKim, Washington, D. C.

Rev. Chas. F. J. Wrigley, M.A., Buffalo, N. Y.

Rev. William Marshall, Coudersport, Pa.

Rev. James Alex. Dickson, Governor's Island, N. Y.

Rev. Charles M. Armstrong, West Philadelphia, Pa.

Rev. William Cooke, Oriskany, N. Y.

Rev. David S. McCaslin, D. D., Minneapolis, Minnesota.

Rev. James N. Chalmers, Lonsdale, R. I.

Rev. Percy Stickney Grant, New York City.

Rev. F. P. Berry, Maryville, Mo.

Rev. Benjamin F. Matrau, A. M., Chicago, Ill. :

"Vivisection degrades — superinduces brutality and indifference to pain of others."

Rev. Wm. A. McCorker, Detroit, Mich. : —

"Unless vivisection is *absolutely necessary* to protect human life and to guard against human suffering, I am perfectly clear that it should not be practised. Whether there is ever such necessity I hardly feel competent to decide. If there ever is such occasion for its practice, it should be rendered painless. I think my place is with those absolutely prohibiting it."

Rev. Wm. C. Pond, San Francisco, Cal.

Rev. Washington R. Laird, West Chester, Pa.

Rev. Lester L. West, D.D., Winona, Minn.

Rev. Elwell O. Mead, A. M., Burton, Ohio.

Rev. J. C. Ely, Xenia, Ohio.

Rev. Elisha Gifford, Cambridge, Mass. : —

"We believe that animals have rights and should be protected from cruelty : the inhumanity and barbarism of vivisection seem certain, its good effects are doubtful ; the first position therefore seems to us the correct one, that the practice should be abolished, under pains and penalties."

Rev. Albert F. Pierce, Danbury, Conn. : —

"I cannot indorse all in the statement made above. My opinion is that no such benefit results from vivisection as warrants the practice of it; therefore my judgment as well as my sympathies are strongly against it. *If it can be shown* that great good *has* resulted to human life either in the treatment of or prevention of disease, which good could not have come through other methods of investigation, then I should favor the 'second' statement, but under the most stringent regulations. But it will take a great deal more evidence than I have yet seen to produce such a conviction in my own mind. For this reason I am strongly opposed to vivisection in any form."

Rev. F. A. Warfield, Brockton, Mass.

Rev. J. T. Sunderland, A. M., Ann Arbor, Michigan.

Mrs. Eliza R. Sunderland, Teacher of History, Ann Arbor, Mich.

Rev. John H. Morison, D. D., Boston, Mass. :

"I should prefer to sign No. II. were it not that, when once engaged in vivisection without pain, the operator is very liable to be carried beyond his expectations."

Rev. F. C. Butler, Quincy, Mass.

Rev. J. M. G. Foster, Bangor, Maine.

Rev. J. F. Loba, Evanston, Ill.

Rev. Wm. M. Jones, St. Louis, Mo.

Rev. D. H. Temple, Los Gatos, Cal.

Rev. J. Edw. Rielly, Hancock, Mich.

Rev. David Stuart Hamilton, Columbia, Pa.

Rev. Edwin Pond Parker, Hartford, Conn.

Rev. Jos. S. Jenckes, Indianapolis, Ind.

Rev. Daniel L. Furber, Newton Centre, Mass

Rev. Thomas H. Cocroft, Providence, R. I.

Rev. Edward C. Porter. Watertown, Mass.

Rev. Edward Norton, Quincy, Mass.

Rev. S. Wright Butler, D. D., Omaha, Neb.

Rev. Albert Buel Vorse, Wellesley Hills.

Rev. P. A. Gleason, Wakeman, Ohio.

Rev. Wm. Wallace, Calumet, Mich.

Rev. Chas. Addison Northup, Norwich, Conn.

Rev. George A. Tewksbury, Concord, Mass.
Rev. George M. Bartol, Lancaster, Mass.
Rev. Dr. Thomas C. Easton, Washington, D. C.
Rev. Asher Anderson, Meriden, Conn.
Rev. J. S. Plumer, Cadiz, Ohio.
Henry S. Clubb, Editor, Philadelphia, Pa.
Rev. W. Moore Jones, M. A., Murphysboro, Ill.
Rev. Arthur W. Spooner, Camden, N. J.
(Erases 1st sentence of 2d paragraph.)
Rev. Dr. Henry B. Cromwell, Brooklyn, N. Y.
Rev. Eugene A. Johnson, Washington, D. C.
Rev. Walter H. Waygood, Schenectady, N. Y.
Rev. William Rounseville Alger, Author, Boston, Mass.
Rev. Newman Hall, D. D., Vine House, Hampstead, London, Eng.
Mark W. Harrington, Chief of Weather Bureau, Washington, D. C.
James Jeffrey Roche, Editor of "The Pilot," Boston, Mass.: —
"I do not believe that even an advocate of 'unrestricted vivisection' should be subjected to the mercy of his fellow vivisectors; but therein, perhaps, I err on the side of tenderness."
Rev. George Hughes, Editor of "The Christian Standard," Philadelphia, Pa.
Rev. James O. S. Huntington, Westminster, Maryland: —
"There seems to be an increasing tendency on the part of people at large to act as though the principal reason for being on this planet were to continue to exist in the life of this world. I believe that the very opposite is true; that man exists in order to die, and through death to find life."
Rev. John A. Bevington, B. D., Barnstable, Mass.
Rev. H. D. Lathrop, A. M., D.D., St. Helena, California.
Rev. Jas. H. Denny, V. P., A. H. A., Rochester, N. Y.
Rev. John T. G. Nichols, S.T. D., Cambridge, Mass.: —
"I heartily indorse the above without change or qualification. I regard vivisection as one of the most monstrous wrongs and cruelties of our times."
Rev. F. R. Farrand, A. B., San Francisco, Cal.
Rev. Myron W. Reed, Denver, Colo.
Rev. Francis J. Clerc, D. D., Philipsburg, Penn.
Rev. J. F. Packard, Author and Editor of "Messiah's Herald," Walnut Hill, Mass.
Rev. W. Simonton, D. D., Emmittsburg, Md.

Rev. George McClellan Fiske, D. D., Providence, R. I.
Rev. Charles D. Bell, D. D., Cheltenham, England.
Clara B. Colby, Editor, Washington, D. C.
P. W. Raidobough, Editor, Chicago, Ill.
J. A. Mitchell, Editor, New York City.
C. A. Bickford, Editor, Boston, Mass.
E. C. Townsend, Editor, New York City.
Henry B. Williams, Bristol, Vt.
Hon. Rowland B. Mahany, M. C., Washington, D. C.
Prof. Geo. L. Collie, Beloit College, Beloit, Wisconsin.
Prof. J. L. Patterson, Union College, Schenectady, N. Y.
Hon. Geo. C. Crowthers, St. Joseph, Mo.
Hon. J. E. Jones, Governor of Nevada.
Hon. J. E. Rickards, Governor of Montana.
Hon. Mat Hoke, Nashville, Tenn.
John T. Dale, Esq., Chicago, Ill.
A. S. Pratt, Pres. Washington Humane Soc., Washington, D. C.
Hon. Rufus Dane, Mobile, Alabama.
Henry D. Lloyd, Editor of the "Chicago Tribune," Chicago, Ill.
B. O. Flower, Editor of "The Arena," Boston, Mass.
Mrs. Robert Treat Paine, Boston, Mass.
Mrs. Grace S. Voorhees, W. Winsted, Conn.:
"After having had an extended course in college in vivisection under a Christian professor."
W. J. Brewster, Hannibal, N. Y.
Henry S. Salt, Author, London.
"Ouida." Author, Florence, Italy.
Chas. H. Grandgent, A. B., Director of Modern Language Instruction in the Boston Public Schools, Cambridge, Mass.
Prof. J. Seymour Slie, A.M., Topeka, Kansas.
Willard J. Hull, Esq., Minneapolis, Minn.
D. O'Loughlin, Editor of the "Twentieth Century," New York City.
Hon. W. A. Peffer, U. S. Senator, Kansas.
Hon. Dennis M. Hurley, M. C., Brooklyn, New York.
Hon. Jove L. Johnson, M. C., Sacramento, California.
Hon. Jacob H. Gallinger, M. D., U. S. Senator, N. H.
Hon. Thomas W. Palmer, Ex-Senator from Michigan.
Hon. Th. McEwan, Jr., Member of Congress, New Jersey.
Hon. Elijah A. Morse, Member of Congress, Mass.
Hon. L. M. Strong, M. C., Ohio.

Hon. J. D. Leighty, M. C., Ind.

Byron B. Northrup, Esq., Racine, Wis.

Charles A. Hamlin, Esq., Syracuse, N. Y.

Christopher Roberts, Esq., Newark, N. J.

R. H. Carothers, Associate Editor of " The Southern School," Louisville, Ken.

James P. Magenis, Esq., Editor of the " Adams Freeman," Adams, Mass.

H. L. Green, Esq., Editor of the " Free Thought Magazine," Chicago, Ill.

"I settle this question in my own mind the same way that Garrison settled the slavery question: 'As I knew it was wrong to sell my children on the auction block, I decided from that it would be wrong to sell any other man's children.' So I say, ' I would allow no cruel treatment on any living creature that I

would not permit on my own children.' "

Prof. James Otis Lincoln, San Mateo, Cal.

Prof. W. C. Esty, Amherst, Mass.

Prof. B. D. Cockrill, Pres. Trinity University, Tehuacana, Texas.

Mrs. Olive Thorne Miller, Bird Student and Author, Brooklyn, Mass.

Miss Louise Imogen Guiney, Writer, Auburndale, Mass.

Mrs. Elizabeth Stuart Phelps Ward, Author, Newton Centre, Mass.

Mr. Herbert D. Ward, Author.

Warren Lee Goss, Author, Norwich, Conn.

Edward Maitland, Esq., Author, London, England.

C. W. Emerson, Emerson College of Oratory, Boston, Mass.

II. VIVISECTION ALLOWABLE IF WITHOUT PAIN.

WHETHER that experimentation upon living animals known as Vivisection is justifiable or not, depends, in our judgment, *exclusively on the question of pain.* Utility alone cannot give Science her authority; for then might she ask, as she sometimes has already done, for the living bodies of the criminal, the idiot, or the savage, wherein to search for the mysteries of life. We believe that Man's dominion over the animated world beneath him is not absolute; that the power to kill does not cover the right to torture; that our power is responsible; that our Humanity invests us with certain moral obligations toward the lower forms of life; and that under these Man is not free to treat even the scorpion or the tiger as they might treat him. But the use of chloroform and ether have made it possible to perform certain experiments and demonstrations upon living animals without the slightest pain, and these only we regard as justifiable for demonstration or research.

The dangers of this practice, however, are so many, the temptations to excess are so strong, the abuses to which it has led are so notorious and deplorable, that the decision of this question of pain should not be left to the judgment of each experimenter; but the whole practice, like the study of human anatomy with dissection, should be regulated by definite laws, confined to certain objects, permitted only to competent and trustworthy persons, and restricted to licensed places which shall be open at all times to inspection by the Presidents of Humane Societies for Protection of Animals or their authorized representatives.

VIVISECTION ALLOWABLE IF WITHOUT PAIN.

ALBERT L. GIHON, A. M., M. D., Medical Director, U. S. Navy, in charge of the U. S. Naval Hospital, Washington, D. C.

Prof. HENRY M. FIELD, M. D., Emeritus Professor of Therapeutics, Dartmouth Medical College : —

"I give the above my emphatic approval. But if vivisection thus restricted and guarded is not attainable, I should affix my signature to No. I."

CLINTON WAGNER, M. D., Senior Surgeon of the Metropolitan Throat Hospital, New York City : —

"Vivisection may be allowable if without pain' and performed by the Professor or his licensed assistants, and only in the laboratories of incorporated medical schools."

W. C. BOUTON, A. B., M. D., Clinical Instructor in Neurology at the Northwestern University Medical School, Chicago, Ill. : —

"I do not believe that vivisection should be wholly dispensed with, for I believe that we can still gain some valuable knowledge from it. But I believe, as stated, that it should be without pain, regulated by definite laws, confined to certain objects, permitted only to competent and trustworthy persons, and restricted to licensed places which shall be open at all times to inspection by the Presidents, or their authorized representatives, of Humane Societies for Protection of Animals."

Prof. H. B. CUMMINS, M. D., Professor of Physiology, Lincoln, Nebraska.

Prof. FRANKLIN TOWNSEND, A. M., M. D., Professor of Physiology, Albany Medical College, Albany, N. Y.

Prof. J. C. HARTZELL, Jr., M. S., Professor of Biology, Claflin University, Orangeburg, S. C.

Prof. J. H. ETHERIDGE, A. M., M. D., Professor of Obstetrics and Gynæcology, Rush Medical College, Chicago, Ill.

Prof. H. D. CHAMPLIN, M. D., Professor of Nervous Diseases, Cleveland University of Medicine and Surgery, Cleveland, Ohio : —

"I do not believe in these cases any tyro should be allowed to vivisect ; nor do I believe in vivisection just to verify old experiments. Unless something of great value is to be gained in a scientific way, it should be forbidden even under the influence of an anæsthetic."

Prof. W. T. WENZELL, M. D., Ph. D., Professor of Chemistry, etc., University of California, San Francisco, Cal.

Rev. FREDERIC R. MARVIN, M. D., Troy, N. Y.: —

"I believe vivisection should be allowable in cases where pain may be avoided, and then only as conducted by experts for some definite end of sufficient consequence. It should never be allowed for mere purposes of demonstration, or as a method of instruction in the class-room or in the medical college."

(In another letter to the Association, Dr. Marvin says : "Though now a minister of the Gospel, I was educated to the profession of medicine and was graduated from the College of Physicians and Surgeons, 'Medical Department of Columbia College, N. Y.,' in 1870. In the class-room I saw vivisections so unqualifiedly cruel that even now they remain in my memory as a nightmare. I am persuaded that none of the so-called experiments upon living animals that I witnessed were of any real value to me or to my fellow-students.")

A. N. BROCKWAY, A. M., M. D., New York City: —

"*My opinion is that no experimenter should inflict pain on any animal which he would not himself be willing to suffer in the same cause.*"

WILLIAM WALLACE GARDNER, M. D., Springfield, Mass.: —

" I believe it useful under proper restrictions to save human suffering. *What I should be willing to suffer voluntarily, the lower order of animals should be obliged to suffer for humanity's sake.*"

W. S. TREMAINE, M. D., Buffalo, N. Y.: —

" Personally, I would rather perform vivisection on human beings than on the lower animals, for the reason that the object can be made intelligent to them; and, again, there is some value to the world both morally and physically. Many of the so-called ' advances ' in surgery are of little or no value, except as advertising dodges for ambitious young surgeons."

HENRY W. SAWTELLE, M. D., Surgeon in the U. S. Marine Hospital, New Orleans, La.

ALBERT H. BLANCHARD, M. D. (Harvard), Sherborn, Mass.: —

" It appears to me that the advantages of vivisection, and the practical good derived therefrom, are not at present at least, sufficient to justify its practice unless it can be done without pain."

C. J. CLEBORNE, M. D., Medical Director, U. S. Navy, Naval Hospital, Norfolk, Va.

Prof. J. HENRY JACKSON, A. M., M. D., Professor of Physiology, Barre, Vermont.

FRANK W. RING, M. D., A. M., Surgeon to Manhattan Eye and Ear Hospital, New York City.

N. A. MOSSMAN, M. D., New York City: —

" Without supervision, indifferent experimenters might say that they had complied with all the requirements if they gave a few inhalations of chloroform, then experimented any length of time without continuing its use."

JOHN WESLEY DAILY, A. B., M. D., Boston, Mass.: —

" Man, by reason of superior intelligence, does not own the earth. He is simply one of the creatures growing out of the conditions that give and govern life throughout the limitless universe of Nature, and in all the countless worlds. Who knows or can even guess the ultimate unfoldments of time ? In the mind of the crowning intelligence that may finally dominate this earth, man may appear as the man-monkey of the Miocene epoch or the anthropoid ape of the present. Until it is shown that vivisection actually benefits the animals, as a class, upon which it is practised, man's right to use his scalpel upon live, quivering animal structures is very questionable."

EDWARD W. AVERY, M. D., Brooklyn, N. Y.: —

"Even though anæsthetics are used, I do not consider vivisection justifiable to demonstrate well-known facts or to gratify curiosity."

EDGAR S. DODGE, M. D., Natick, Mass.: —

"Vivisection should be under the limitation and control of United States law, and severe penalties should be applied to all violations."

CLARKSON C. SCHUYLER, M. D., Plattsburg, N. Y.: —

"If a law allowing vivisection and without pain cannot be enforced, then I am for the absolute prohibition of vivisection."

W. E. Sparrow, M. D., Mattapoisett, Mass.

H. H. Brigham, M. D., Fitchburg, Mass.

James Utley, M. D., Newton, Mass.

E. A. Chase, M. D., Brockton, Mass.

P. H. Keefe, M. D., Worcester, Mass.

M. L. Lindsay, M. D., Athol, Mass.

E. A. Deane, M. D., Montague, Mass.

Geo. E. Foster, M. D., Springfield, Mass.

John B. Chagnon, M. D., Fall River, Mass.

C. B. Bridgham, M. D., Cohasset, Mass.

John Blackmer, M. D., Springfield, Mass.

(Makes slight changes in phraseology, but not to impair the general principle.)

Henry J. Cushing, M. D., Merrimac, Mass.

Luther M. Lee, M. D., Dorchester, Mass.

S. J. Grover, M. D., Brockton, Mass.

N. R. Perkins, M. D., Dorchester, Mass.

Wm. O. Hunt, M. D., Newtonville, Mass.

George W. Doane, M. D., Hyannis, Mass.

M. S. Soule, M. D., Winthrop, Mass.

S. W. Clark, M. D., Lynn, Mass.

F. A. Shurtleff, M. D., Somerset, Mass.

Charles W. Stiles, M. D., Newburyport, Mass.

J. E. Blaisdell, M. D., A. M., Chelsea, Mass.

Francis Collamore, M. D., North Pembroke.

W. G. Eaton, M. D., Lowell, Mass.

C. R. Starkweather, M. D., West Cummington, Mass.

S. T. Davis, M. D., Orleans, Mass.

J. R. Greenleaf, M. D., Gardner, Mass.

Wm. T. Souther, M. D., Worcester, Mass.:

"During my course at Harvard (in 1873), experiments upon animals were performed painlessly; ether was used invariably, even upon frogs and tadpoles."

L. W. Curtis, M. D., Globe Village, Mass.

P. L. Sanborn, M. D., Marblehead, Mass.

Edgar C. Collins, M. D., Springfield, Mass.

Daniel March, Jr., M. D., Winchester, Mass.

T. F. Goodwin, M. D., Mt. Vernon, N. Y.

J. E. Hamill, M. D., Phœnix, N. Y.

D. D. Wickham, M. D., Port Jervis, N. Y.

S. W. Reed, M. D., Morgantown, N. Y.

George N. Wilson, M. D., Boston, Mass.

Charles M. Fuller, M. D., Boston, Mass.

Edwin H. Brigham, M. D., Boston, Mass.

Isaac Farrar, M. D., Boston, Mass.

J. Foster Bush, M. D., Boston, Mass.

A. W. K. Newton, M. D., Boston, Mass.

James Louis Beyea, M. D., New York City.

Egbert H. Grandin, M. D., New York City:

"Vivisection is allowable if without pain, except when, in the judgment of recognized experts, new facts of value to the human race cannot be certified if the animal be anæsthetized."

W. H. Vermilye, M. D., New York City.

Dayton W. Searle, A. M., M. D., New York City.

G. S. Cook, M. D., New York City.

John Pitkin, M. D., Buffalo, N. Y.

Alcinous B. Jamison, M. D., New York City.

John I. Brinkerhoff, M. D., Auburn, N. Y.

Alexander Berghaus, M. D., New York City.

Richard E. Kunze, M. D., New York City.

Frederic J. Kneuper, M. D., New York City.

John G. Linsley, M. D., New York City: —

"I accept No. III. in regard to use of drugs, and No. II. in regard to vivisection per se. Nos. II. and III. combined will express my views upon the subject."

James G. Birch, M. D. (Harvard Medical School), Newburgh, N. Y.

Edward G. Tripp, M. D., New York City.

Geo. W. Brush, M. D., Brooklyn, N. Y.

A. Henry Hart, M. D., Brooklyn, N. Y.

William O'Meagher, M. D., New York City.

M. K. Vedder, M. D. College Physician & Surgeon, New York City.

(Dr. Vedder underscores last three lines.)

A. D. Rockwell, A. M., M. D., New York City: —

"Vivisection is allowable if without pain and restricted by utility."

Edward F. Quinlan, M. D., New York City:

"I am radically opposed to vivisection *without* any restrictions, as also am I opposed to the absolute prohibition of vivisection. . . . Vivisection allowable without pain meets with my hearty indorsement."

Martin Burke, M. D. (Bellevue), New York City: —

"This is a very proper statement."

Charles Carter Cranmer, M. D., Surgeon, New York City.

A. D. Tewksbury, M. D., Elmira, N. Y.

M. A. Southworth, M. D., Little Falls, N. Y.

W. F. Sanford, M. D., Webster, Mass.

E. F. Spaulding, M. D., Boston.

Geo. J. Moser, M. D., New York City: —

"The animal experimented on should not come out of the influence of anæsthetics after an important fact has been demonstrated, or a major operation has been performed."

D. N. Barker, M. D., Broadalbin, Fulton Co., N. Y.: —

"I indorse as above if regulated by definite State laws, permitted only to competent persons," etc.

J. P. Wheeler, M. D., Brighton, N. Y.

O. A. Jakway, M. D., Brockport, N. Y.

A. B. Sloan, M. D., Belloner, N. Y.

M. Talbot, M. D., Niagara Falls, N. Y.

J. H. Westcott, M. D, Norwich, N. Y.

J. C. Benham, M. D., Hudson, N. Y.

H. P. Vosburgh, M. D.. Halsey Valley, N. Y.

John H. Fitch, M. D., New Scotland, N. Y.

S. O. Gleason, M. D., Elmira, N. Y.

John H. Mitchell, M. D., Newburgh, N. Y.

H. C. Johnston, M. D., New Brighton, N. Y.

J. H. Copp, M. D., Natural Bridge, N. Y.

E. H. Barnes, M. D., Marathon, N. Y.

Walter E. Lauderdale, M. D., Geneseo, N. Y.

E. D. Coonley, M. D., Port Richmond, N. Y.

Jas. C. Spiegel, M. D., Middletown, N. Y.

C. B. Warner, M. D., Port Henry, N. Y.

H. C. Miller, M. D., Greenbush, N. Y.: —

"In my opinion vivisection should be performed on criminals condemned to death."

Samuel W. Abbott, M. D., Wakefield, Mass.:

"Vivisection is allowable, if without unnecessary or excessive pain."

John E. Losee, M. D., Red Hook, N. Y.

R.C.VanWyck,M.D.,Hopewell Junction,N.Y.

E. M. Draper, M. D., Ilion, N. Y.

O. A. Bruce, M. D., Hyndsville, N. Y.

Alonson Bishop, M. D., Ithaca, N. Y.

C. Spencer Kinney, M. D., Middletown, N. Y.

Lucien L. Brainard, M. D., Little Falls, N. Y.

Wm. M. James, M. D., Whitesboro, N. Y.

Otto Risch, M. D., Brooklyn, N. Y.

David A. Gordon, M. D. Brooklyn, N. Y.

John C. MacEvitt, M. D., Brooklyn, N. Y.

O. C. Stout, M. D., Syracuse, N. Y.

Frank H. Green, M. D., Homer, N. Y.

J. Russell Taber, M. D., Brooklyn, N. Y.

Z. Brooks Wales, M. D., Elmira, N. Y.

Theron A. Wales, M. D., Elmira, N. Y.

A. M. Larkin, M. D., Norwood, N. Y.

John W. Benton, M. D., Ogdensburg, N. Y.

John M. Julian, M. D., Pleasant Valley, N. Y.

Lyman Barton, M. D., A. M., Willsboro, N.Y.

Cassius J. Logans, M.D., Warrensburgh, N.Y.

T. James Owens, M. D., Steuben, N. Y.

G. H. Lathrop, M.D., Livingston Manor, N. Y.

T. J. Green, M. D., Mexico, N. Y.

J. H. Helmer, M. D., Lockport, N. Y.

J. H. Trumbull, M. D., Hornellsville, N. Y.

Ira D. Hopkins, A. M., M. D., Utica, N. Y.

William L. Harding, M. D., New York City.

Lyman A. Clark, M. D., Cambridge, N. Y.

J. R. Brown, M. D., Seward, N. Y.

Nelson W. Bates, M.D., Central Square, N.Y.

A. Von Der Luhe, M. D., Brooklyn, N. Y.

Edward B. Foote, M. D., New York City.

Lyman Watkins, M. D., Brooklyn, N. Y.

H. C. Sutton, M. D., Rome, N. Y.

L. M. Johnson, M. D., Greene, N. Y.: —

"Except in some very rare cases, when the use of chloroform or ether is inadmissible, or rather utterly *impossible*, with a due regard to *necessary* scientific experimentation."

Henry B. Burton, M. D., Troy, N. Y.

George M. Lamb, M. D., Hoosick Falls, N. Y.

Samuel Blume, M. D., Riverhead, L. I., N. Y.

Elliott C. Howe, M. D., Lansingburgh, N. Y.

Henry W. Caldwell, M. D., Pulaski, N. Y.:

"I fully concur with and indorse the above as the only right we have for official demonstrations upon the living."

Jos. Alfred Deane, M. D.. Catskill, N. Y.

A. J. Alleman, M. D., MacDougall, N. Y.

Benj. C. Wakely, M. D., Hornellsville, N. Y.

J. Denniston, M. D., Ovid, N. Y.

Reuben S. Myers, M.D., Clarence Center, N.Y.
S. W. Wetmore, M. D., Buffalo, N. Y.
Mary B. Wetmore, Buffalo, N. Y.
Jerome Angel, M. D., Cortland, N. Y.
R. E. Belding, M. D., Troy, N. Y.
Mary E. Bond, M. D., New York City.
II. L. Chase, M. D., Palmyra, N. Y.
C. S. Boyce, M. D., Salamanca, N. Y.
Henry A. Jewett, M. D., Northborough, Mass.
Wm. Hagadorn, M. D., Gilboa, N. Y.
Achilles Rose, M. D., New York City.
A. P. Farries, M. D., Florida, N. Y.
C. O. Johnson, M. D., Gowanda, N. Y.
F. W. Advance, M. D., Elmira, N. Y.
J. A. Westlake, M. D., Elmira, N. Y.
Silas Pinckney Holbrook, M. D., East Douglass, Mass.
David F. Atwater, M. D., Springfield, Mass.
Edward H. Ellis, M. D., Marlboro, Mass.
Hiram B. Cross, M. D., Jamaica Plain, Mass.
H. A. Fiske, M. D., East Longmeadow, Mass.
M. Bonner Flinn, M. D., Worcester, Mass.

II. Warren White, M. D., Roxbury, Mass. (Physician to the St. Elizabeth and Baptist Hospitals).
Charles P. Morrill, M. D., North Andover Depot, Mass.
f. W. Whittemore, M. D., Cambridge, Mass.
Francis L. Babcock, M. D, Dedham, Mass.
Edward P. Scales, M. D., Newton, Mass.
Charles L. French, M. D., Clinton, Mass.
E. D. Hutchinson, M. D., Westfield, Mass.
James R. Deane, M. D., Newton Highlands, Mass.
Alvin M. Cushing, M. D., Springfield, Mass.
E. W. Higher, M. D., Northampton, Mass.
Asa V. Snow, M. D., Brookfield, Mass.
Albert B. Robinson, M. D., Boston, Mass.
J. F. Shurtleff, M. D., South Middleborough, Mass.
John Langdon Sullivan, M. D., Malden, Mass.
E. P. Hussey, M. D., Buffalo, N. Y.
Clayton L. Hill, M. D., Buffalo, N. Y.
J. S. Helbert, M. D., Buffalo, N. Y.
II. A. Morse, M. D., Batavia, N. Y.

Mr. WILLIAM DEAN HOWELLS, Author, New York City.

Mr. EDWARD BELLAMY, Author, Chicopee Falls, Mass.

Mr. BRANDER MATTHEWS, Author, New York City.

Prof. ALBERT BUSHNELL HART, Harvard College, Cambridge, Mass.

Prof. JOHN BASCOM, Williamstown, Mass.

Prof. ALBION W. SMALL, Ph.D., Professor of Sociology, University of Chicago.

Prof. JOHN GRIER HIBBEN, Professor of Logic, Princeton University, Princeton, N. J.

Prof. CHARLES W. SHIELDS, Princeton, N. J.: —

"Vivisection is allowable if without pain, in the judgment of humane and scientific experts."

Prof. W. F. HEWETT, Cornell University, Ithaca, N. Y.

Prof. GALUSHA ANDERSON, D. D., LL.D., Professor of Practical Theology, University of Chicago.

Prof. H. S. WHITE, Dean of Cornell University, Ithaca, N. Y.:

"Vivisection is allowable if for true scientific purposes, in order ultimately to ameliorate the condition of mankind, as well as of animal life in general."

Prof. S. BURNHAM, Dean of the Hamilton Theological Seminary, Hamilton, Madison Co., N. Y.

Prof. HENRY K. EDSON, Grinnell, Iowa.

Prof. W. G. TOUSEY, A. M., B. D., Tufts College, Mass.

Prof. M. L. D'OOGE, University of Michigan, Ann Arbor, Mich.

Prof. GABRIEL CAMPBELL, M. P., S. T. D., Dartmouth College, Hanover, N. H.

Prof. GEORGE E. WOODBERRY, Columbia College, New York City.

(Erases all but last paragraph, and inserts "Public Officers" in last line but one.)

Prof. JAMES H. HYSLOP, Columbia College, New York City.

Prof. CHARLES C. ROUNDS, Ph. D., Principal of the State Normal School, Plymouth, N. H. : —

"Vivisection is allowable if without pain, for research, — not for illustration and demonstration."

President E. BENJAMIN ANDREWS, D. D., LL.D., Brown University, Providence, R. I. : —

"I would add that, if there are results of very great importance obtainable by vivisection without anæsthesia or with limited anæsthesia, I would permit vivisection without or with limited anæsthesia subject to the conditions recited above in the last paragraph of printed matter."

President ANDREW V. RAYMOND, LL.D., Union College, Schenectady, N. Y.

President JAMES R. DAY, D. D., S. T. D., Chancellor of the Syracuse University, Syracuse, N. Y.

President GEORGE A. GATES, D. D., Iowa College, Grinnell, Iowa :

"This, on the whole, seems nearest my own view. There are perhaps some cases where pain may be an essential factor of the investigation. This ought to be of the shortest possible duration consistently with the scientific purpose. More depends at last upon the right kind of heart in the bosom of the operator."

President CHAS. F. MESEROL, A. M., Shaw University, Raleigh, N. C.

President M. WAHLSTROM, A. M., Ph.D., Gustavus Adolphus College, St. Peter, Minn.

President CARL A. SWENSSON, Bethany College, Lindsborg, Kansas : —

"Vivisection is allowable if without pain, and for object stated in No. III."

President FENTON GALL, Hillsboro College, Hillsboro, Ohio.

President WM. H. PURNELL, A. M., LL. D., New Windsor College, New Windsor, Maryland : —

"The fourth statement is horrible ; the third may be tolerated ; the second expresses my opinion."

Mr. Robert S. Davis, Editor of " The Call," Philadelphia, Pa.

Mrs. Frank Leslie, Author and Publisher, New York City.

Prof. James Swan Barrell, A. M., Master of Harvard School, Cambridgeport, Mass.

Dr. H. Lansing, Editor, New York City : —
"Vivisection allowable if without pain, and restricted by utility."

Mr. Henry B. Blackwell, Editor of " Woman's Journal," Boston, Mass.

Rt. Rev. JOHN JAMES STEWART, Bishop of Worcester, Worcestershire, England : —

"I think II. and III. may be combined. Vivisection I think is allowable when restricted by utility, and when it is performed without pain to the animal operated upon."

Rt. Rev. THOMAS A. JAGGER, Bishop of Southern Ohio.

Rt. Rev. GEORGE F. SEYMOUR, Bishop of Springfield, Ill.

Rt. Rev. DANIEL S. TUTTLE, Bishop of Missouri, St. Louis, Mo.

Rt. Rev. C. K. NELSON, Bishop of Georgia, Atlanta, Ga.

Rt. Rev. O. W. WHITAKER, Bishop of Pennsylvania.

Rt. Rev. FRANCIS K. BROOKS, Bishop of Oklahoma and Indian Territories, Guthrie, Oklahoma.

Rev. C. W. LEFFINGWELL, A. M., D. D., Editor of "The Living Church," Chicago, Ill. : —

"To be allowed only for instruction of students in medicine and surgery."

Rev. GEORGE J. MAGILL, D. D., Newport, R. I. : — ·

"The experiments should be such as to inflict no pain during the operation and to leave the subject in a painless condition afterwards."

Rev. WILLIAM P. SHROM, D. D., Pittsburg, Pa. : —

"I should much prefer to sign No. I., but am not quite sure but it would hinder to some extent the furthering of scientific investigation. My whole heart is on that side of the question, and my head more than half there too; yet for the present I give the *investigator* the benefit of my doubt."

Rev. HERMAN C. RIGGS, D. D., Rochester, N. Y. : —

"No. II. comes nearest of your four forms to being my answer. But I favor this limited permission only with the condition that experimentation shall be conducted by none but expert investigators, and by them only for *new* or unsettled experiments. A result once definitely and clearly reached by authoritative investigation, vivisection with reference to that should immediately cease to be permissible."

Rev. ROBERT AIKMAN, D. D., Madison, N. J. : —

"It seems to me that the above statement meets the demands of an enlightened and humane public opinion. The pain of instantaneously inflicted death is not greater than the ordinary pains of natural death, and death under the influence of chloroform is without pain. There seem to have been discoveries made of the causation of disease and of remedial agencies in the relief and cure of diseases which make painless vivisection allowable; and especially if the places where it is performed are open to inspection as above."

Rev. John Hall, D. D., New York City.

Rev. Charles H. Eaton, D. D., New York City.

Rev. Reuen Thomas, D D., Brookline, Mass.

Rev. H. Martyn Hart, D. D. (Dean of St. John's Cathedral), Denver, Colo.

Rev. Granville W. Nims, Walton, N. Y.

Rev. Joseph Anderson, D. D., Waterbury, Conn.

Rev. William MacConnock, M. A., St. Ann's Church, Brooklyn, N. Y.

Rev. Joseph Henry Allen, D. D., Cambridge, Mass.: —
"The whole thing is so repulsive and abhorrent to me that I cannot imagine instances in which my *personal conduct* would not be controlled by your first condition of 'absolute prohibition.' But something must be conceded by a layman to experts; and I think the conditions expressed above (No. II.) is all that can be attempted wisely by general law. And I do not feel justified in saying that a scrupulously medical specialist in an extreme case may not conform with No. III.; not, however, the ordinary lecturer or practitioner."

Rev. Henry E. Niles, D. D., York, Pa.

Rev. W. C. Gannett, Rochester, N. Y.: —
"Vivisection allowable if without pain, but only for purposes of investigation (never of illustration), and only to persons duly licensed under stringent conditions of competency, place, time, and inspection; all other vivisections to be a misdemeanor punishable by imprisonment (not fine)."

Mrs. Isabel C. Barrows, M. D., Editor, Boston, Mass.

Rev. Samuel J. Barrows, Editor of the "Christian Register," Boston, Mass.

Rev. A. J. Lyman, D. D., Brooklyn, N. Y.

Rev. Wm. A. Vibbert, S. T. D., New York City: —
"Vivisection allowable if without pain and restricted by utility."

Rev. W. W. Moorhead, D.D., Greensburg, Pa.
Rev. Edward C. Hood, Wrentham, Mass.
Rev. Charles H. Oliphant, Methuen, Mass.
Rev. W. M. Backus, Alton, Ill.: —
"I am not well enough versed upon this subject to express an opinion intelligently, but the above seems to me to be the rational and humane conclusion."

Rev. J. W. Bailey, Lockport, N. Y.
Rev. Thomas P. Byrnes, Humboldt, Iowa.
Rev. S. A. Hoyt, D. D., Watertown, N. Y.
Rev. Geo. W. Wood, Mt. Morris, N. Y.
Rev. Mary H. Graves, Boston, Mass.
Rev. John H. Munro, D. D., Philadelphia, Pa.
Rev. W. S. Fulton, D. D., Lexington, Ky.
Rev. A. B. Carver, Yonkers, N. Y.
Rev. Geo. H. Ide, Milwaukee, Wis.
Rev. Alvin F. Bailey, Barre, Mass.: —
"Vivisection allowable if without pain, and restricted by utility."

Rev. Theodore S. Rumney, Germantown, Pa.

Rev. John Townsend, M.A., Middletown, Ct.
Rev. Richmond Shreve, D.D., Albany, N.Y.:
"Vivisection *justified by utility and painlessness;* to be restricted by law to certain definite objects, and surrounded by every possible safeguard against license or abuse."

Rev. D. O. Mears, Cleveland, Ohio.
Rev. Lemuel B. Bissell, Monroe, Mich.
Rev. J. A. Rondthaler, Indianapolis, Ind.:
"If the first paragraph under 'Absolute Prohibition of Vivisection' could be proved to be absolutely true, I would sign that number. However, there are as many high in knowledge and authority that express opposite views to those cited there that I sign number II. in the earnest hope that its provisions may obtain."

Rev. Edward H. Krans, New York City.
Rev. John T. Huntington, Hartford, Conn.
Rev. John N. Lewis, Jr., B. A., Honesdale, Pa.
Rev. Wm. C. Richardson, Newburyport, Mass.
Rev. Wm. P. Orrick, D. D., Reading, Pa.
Rev. Henry M. Ladd, D. D., Cleveland, Ohio.
Rev. J. Sanders Reed, Watertown, N. Y.
Rev. Charles James Wood, S. T. B., York, Pa.
Rev. Marcus A. Tolman, Mauch Chunk, Pa.
Rev. J. Frederick Sexton, M. A., Cheshire, Ct.
Rev. Dr. Henry McCrea, New Haven, Conn.:
"In absence of accurate knowledge of the facts, I endorse the 1st or 2d propositions in your circular, strongly inclining to the 1st."

Rev. Frederick B. Allen, Boston, Mass.
Rev. James H. McIlvaine, New York City.
Rev. H. W. Nelson, Jr., D. D., Geneva, N. Y.
Rev. C. S. Brooks, Fitchburg, Mass.
Rev. A. Z. Conrad, Worcester, Mass.
Rev. George H. Reed, Haverhill, Mass.
Rev. Charles E. Jefferson, B. A., S. T. B., Chelsea, Mass.
Rev. Louis Gregory, Lincoln, Neb.
Rev. Smith Baker, D. D., East Boston, Mass.
Rev. S. S. Mitchell, Buffalo, N. Y.: —
"Am in doubt between this and absolute prohibition."

Rev. A. W. Ringland, D. D., Toledo, Ohio.
Rev. C. H. Tyndall, M. A., New York City.
Rev. James Henry Wiggin, Boston, Mass.
Rev. Thomas M. Miles, Bristol, Conn.
Rev. Dr. Geo. K. Ward, Dansville, N. Y.
Rev. James Dingwell, Rockville, Conn.
Rev. Nelson Millard, D. D., Rochester, N. Y.
Rev. Alexander McKenzie, Cambridge, Mass.
Rev. J. Nelson Trask, New Salem, Mass.

Rev. Charles Olmstead, Cambridgeport, Mass.:
" Should prefer these points softened a little
for *cases of rare emergency.*"

Rev. T. S. Hamlin, Washington, D. C. : —
" I do not accept this statement unqualifiedly. Pain in animals may be justifiable if certain to give knowledge
that will relieve human pain. I would,
therefore, partially approve III. But the
peril is that science will disregard pain
in its zeal of discovery. I favor very
rigid laws of restriction, and every effort
to create a humane public sentiment."

Rev. Wm. G. Poor, B. A., B. D., Keene, N. H.:
" Possibly I would sign the first on a careful
examination of the merits of the case."

Rev. Wm. H. Hudnut, Port Jervis, N. Y.
Rev. Daniel R. Foster, A. M., Trenton, N. J. :
. " I am exceedingly grateful for the privilege
of contributing in the very slightest
towards restricting this evil. Pray do
not become discouraged. It is deeply
rooted, and has concealed environ-
ments."

Rev. Charles Martin Niles, Rutland, Vt.
Rev. Wm. V. W. Davis, Pittsfield, Mass.
Rev. C. S. Richardson, Little Falls, N. Y.
Rev. Heber H. Beadle, Bridgeton, N. J.
Rev. O. W. Folsom, Bath, Me.
Rev. Dan. F. Bradley, Grand Rapids, Mich.
Rev. Chas. H. Bixby, Chicago, Ill.
Rev. Herbert G. Coddington, Syracuse, N. Y.
Rev. J. M. McGrath, Morgan Park, Ill.
Rev. C. H. Hibbard, D. D., Burlington, N. J.
Rev. Edwin S. Gould, Athol Centre.
Rev. Wm. Heakes, Wellsboro, Pa.
Rev. H. P. Dewey, Concord, N. H.
Rev. Frederick J. Bassett, Providence, R. I.
Rev. David S. Schaff, Jacksonville, Ill.
(Dr. Schaff makes numerous slight altera-
tions, but accepts the general statement
as a whole.)

Rev. D. M. Fisk, Toledo, Ohio.
Rev. Wallace Radcliffe, D. D., Detroit, Mich.
Rev. George C. Foley, Williamsport, Pa.
Rev. Francis Edward Smiley, Denver, Colo.
Rev. James B. Nisbett, Brooklyn, N. Y.
Rev. T. Eaton Clapp, Manchester, N. H.

Rev. James R. Winchester, D. D., B. Ph.,
Nashville, Tenn. : —
" My convictions almost take in the first (I.)
form. Man has dominion over brute
creation, and sometimes benefits may
come through a special experiment (of
a painless character). Humane experts
must decide *that painless case.*"

Rev. Henry M. Tenney, Oberlin, Ohio : —
" Exceptions to this rule should be limited
by definite law to experiments essential
to the interests of humanity, and which
are impossible with the use of chloroform or ether, if such there are."

Rev. Wm. H. Scudder, Norwich, N. Y. : —
" Vivisection allowable if without pain, and
restricted by utility."

Rev. James Wm. Ashton, D. D., Olean, N. Y.:
" The above expresses most nearly my view,
but should modify it in exceptional
cases by the following one; namely,
Vivisection restricted by utility, for the
same reason that we observe the principle of utility in dealing with similar
exigencies in the human subject."

Rev. Willis E. Parsons, Danville, Ill.
Rev. Wm. D. Williams, San Francisco, Cal.
Rev. J. L. Countermine, Marinette, Wis.
Rev. David D. Marsh, Unionville, Conn.
Rev. C. E. Dickinson, Marietta, Ohio : —
" I can easily see that vivisection is valuable
to science; at the same time, God has
given nerves to animals as he has to us.
They suffer pain as we do, and I conceive we have no right to inflict upon
them needless pain. Since we can use
anæsthetics with animals as well as with
man, it is our duty to do so. I would
therefore most heartily vote for your
second proposition."

Rev. Marc St. Darling, Sioux City, Iowa.
Rev. O. S. Bunting, Trenton, N. J.
Rev. S. P. Dunlap, M. A., Springfield, Ohio.
Rev. W. F. Murkwick, Ansonia, Conn.
Rev. Perry Wayland Sinks, Painesville, Ohio:
" I find myself in substantial accord with
the second view, with a leaning toward
the first."

Rev. F. P. Davenport, Memphis, Tenn.

President ORELLO CONE, D. D., Buchtel College, Akron, Ohio.
President S. G. GILBREATH, B. Sc., Hiwassee College, Tenn.
President JERE. MOORE, D. D., Tusculum, Tenn. : —
" Vivisection allowable under law."

President SAMUEL PLANTZ, Lawrence University, Appleton, Wis.

President J. D. SHIREY, A. M., North Carolina College, Mt. Pleasant, N. C. : —

"In the main I agree with the above. It expresses my sentiments more nearly than either of the others."

President LEMUEL H. MURLIN, A. B., Baker University, Kansas.

President D. R. DUNGAN, A. M., LL.D., Cotner University, Lincoln, Neb.

President JOHN H. FINLEY, Ph.D., Knox College, Ill.

JAMES W. MOREY, A. M., Lakewood Heights School, N. J.

F. B. KNAPP, S. B., Powder Point School, Duxbury, Mass., President of the Board of Trustees of Partridge Academy : —

"I do not wholly agree with this statement. I am sure that I am opposed to vivisection under nearly all conditions, and am not sure that I would approve it in any, but might under very exceptional circumstances, and so do not quite agree with No. I."

Mr. WM. C. COLLAR, Teacher, Roxbury, Mass.

President W. G. COMPHER, A.M., Scio College, Scio, Ohio.

President W. M. GRIER, D. D., Erskine College, Due West, S. C.

President HOLMES DYSINGER, D. D., Carthage College, Carthage, Ill.

President J. W. BISSELL, D. D., Upper Iowa University, Fayette, Iowa.

President J. J. MILLS, A. M., LL.D., Earlham College, Richmond, Indiana.

President JAMES ROGERS, S. C. & C., A. M., St. Joseph's College, Cincinnati, Ohio : —

"Vivisection allowable if without pain, and restricted by utility."

President B. W. BAKER, M. A., Ph.D., Chaddock College, Quincy, Illinois.

President H. W. McKNIGHT, D. D., LL.D., Pennsylvania College, Gettysburg, Pa. : —

"I object to the repetition of experiments long since made, simply for the entertainment of classes. I would limit it to new experiments, and then only with painlessness and for the highest ends."

Prof. THOMAS R. BAKER, Rollins College, Winter Park, Fla.

Hon. W. H. UPHAM, Governor of Wisconsin, Madison, Wis.

Hon. James M. Brown, Attorney-at-Law. President of the Toledo Humane Society, Toledo, Ohio.

Hon. John Day Smith, Minneapolis, Minn.

Richard F. Reed, Natchez, Miss. : —

"Our dominion over animals carries with it a great trust. We are to protect them in their rights, chief among which is freedom from torture. I am opposed to vivisection for demonstration; and unless the benefit to mankind from it is very decided, I am opposed to it absolutely."

Mr. Chas. R. Fraser, Canton, Ohio.

Mrs. Lily Lord Tifft, Buffalo, N. Y.

Hon. W. A. Calderhead, M. C., Attorney-at-Law, Marysville, Kansas : —
"Nothing but the highest moral purpose can give science her authority ; and when the sense of moral responsibility leaves the operator, every experiment in science becomes dangerous to man. The third statement modified by this II. expresses my views."

Hon. James H. Kyle, U. S. Senator, Aberdeen, So. Dakota: —
"This most nearly accords with my views. In the interests of medical science I think experiments should be made; but I abhor torture in every form."

Hon. Robert Neill, M. C., Batesville, Ark.
Hon. Richard Bartholdt, M. C., St. Louis, Mo.
Hon. W. W. Bowers, M. C., San Diego, Cal.
Hon. J. S. Willis, M. C., Milford, Del.
Hon. Benson Wood, M. C., Effingham, Ill.
Hon. C. J. Boatner, M C., Monroe, La.: —
"Barbarism can hardly be conducive to investigation or the promotion of science, and if it is, the price paid would be too dear. Unless the animal is protected in some way from the torture necessarily inflicted by the operation, it should not be permitted."

Hon. C. D. Clark, Evanston, Mo.
Mr. Philip G. Low, 307 Lenox Ave., New York City.
Hon. Jas. H. Huling, M. C., Charleston, W. Va.
Hon. Walter Evans, M C., Louisville, Ky.:
"Vivisection allowable if without pain, at the same time likely to lead to useful results."

Hon. S. W. Kerr, M. C , Mansfield, Ohio.
Sir Eizak Pitman, Bath, England.

Hon. Robert J. Tracewell, M. C., A. M., Corydon, Md.: —
"I would qualify above as follows: (1) If there was a doubt as to whether any particular operation was painful, I would resolve the doubt in favor of the animal, and prohibit this operation by law. (2) The operation should not be performed on the higher order of animals, such as monkey, horse, or dog."

Mr. A. J. Rowland, Sec. Am. Bap. Pub. Soc., 1420 Chestnut St., Phila., Pa.
Mr. David H. Moore, Editor of the "Western Christian Advocate," Cincinnati, Ohio.
Mr. Charles W. Lincoln, Editor of "The Press," Philadelphia, Pa.
Rev. Thomas Drumm, M. D., 22 State Street, New York City.
Mr. John W. Freese, A. M., Teacher, Cambridge, Mass.
Mr. Frederick S. Cutter, Teacher, Cambridge, Mass.
Miss Caroline T. Haven, Prin. Kin. Dep't., Working-Man's School, New York City.
Mr. Charles W. Birtwell, Gen. Sec. Boston Children's Aid Soc., Boston, Mass.: —
"I should disapprove of vivisection of any kind for purposes of demonstration, and should approve only of painless vivisection by original investigators under State appointment and regulation. I should favor absolute prohibition of vivisection if I did not give the benefit of the doubt to the claim that vivisection may benefit mankind."

Mrs. Charles G. Ames, Boston, Mass.
Rev. Charles E. Woodcock, Christ Church, Ansonia, Conn.
Rev. W. Tatlock, D. D., Stamford, Conn.

III. VIVISECTION RESTRICTED BY UTILITY.

VIVISECTION is a practice of such variety and complexity, that, like warfare between nations, one can neither condemn it nor approve it unless some careful distinctions be first laid down. We hold that only a great and definite advantage to the interests of humanity can justify its use, and that in each case Science must prove that advantage and that necessity; its hands should not be left free to inflict torture without restriction or restraint. Even the zeal of a Magendie or a Mantegazza cannot condone their cruelty, nor can Science make the search for a fact obliterate the distinctions between right and wrong. Within certain limitations we regard Vivisection to be so justified by utility as to be legitimate, expedient, and right. Beyond these boundaries it is cruel, monstrous, and wrong.

Experimentation upon living animals we consider justifiable when employed to determine the action of new remedies; for tests of suspected poisons; for the study of new methods of surgical procedure, or in the search for the causation of disease, — in short, for any object where the probable benefi to mankind is very great, and the suffering inflicted not greater than that of instantaneous death, nor more than the pain and distress of the human ailments to alleviate which the experiment is made. On the other hand, we regard as cruel and wrong the infliction of torment upon animals in the search for physiological facts which have no conceivable relation to the treatment of human diseases; or experiments that seem to be made only for the purpose of gratifying a heartless curiosity, — such, for example, as those described in the work of Professor Mantegazza, entitled "The Effect of Pain upon Respiration."

We consider as *wholly unjustifiable* the common practice in the United States of subjecting animals to torture in the laboratory or classroom, merely for the purpose of demonstrating well-known and accepted facts. We hold that the infliction of torment upon a living animal under such circumstances is not justified by necessity, nor is it a fitting exhibition for the contemplation of youth. And since in England, Scotland, and Ireland such experiments as these are regarded as degrading in tendency, and are therefore forbidden by law, we think no harm will come to Science if they shall also be forbidden in every American State.

We believe, therefore, that the common interests of Humanity and Science demand that Vivisection, like the study of human anatomy in the dissecting-room, should be brought under the direct supervision and control of the State. The practice, whether in public or in private, should be restricted by law to certain definite· objects, and surrounded by every possible safeguard against license or abuse.

[Several shades of opinion are represented by the signatures to this statement. As a rule, disagreement is indicated by various erasures of words, sentences, or paragraphs, made in order to shape the phraseology of the statement into accord with individual views. The signers of this statement may be classified as follows : —

1. Those who have signed the statement without changing it in any way. Many of these have even underscored certain sentences, particularly in the last paragraph.

2. Those who agree in the condemnation of torture as a method of teaching well-known physiological facts, and in general approval of State supervision and control of Vivisection; but who, for various reasons, prefer to soften and change the phraseology of the statement in less important particulars. Their eliminations sometimes refer to single words or short phrases, such as " monstrous," " wholly unjustifiable," " very great," " the common practice," " must prove necessity," " a great advantage," etc. ; or they may affect entire sentences, such as the attempt to define the limits of permissible pain-infliction, the allusion to Magendie and Mantagazza, and particularly the reference to the example of Great Britain. (This last clause is to many especially obnoxious.)

3. Those who denounce or condemn the use of torture as a method of teaching well-known facts, but who cannot approve of any appeal to legislation for its prohibition. As a rule, those who take this position would place no impediments in the way of any original research in any direction. Disapproval of minor points in the phraseology of the statement is quite common.

4. Those who believe that Vivisection should be under more or less supervision by the State, but who apparently would not condemn even painful experiments for teaching purposes, if in the judgment of the teacher pain can be thus made " useful."

5. Those whose erasures, changes, and eliminations affect absolutely and vitally every important part of the entire statement. They do not approve of supervision nor restraint, nor do they condemn any form of experimentation for any object. It is not easy to perceive wherein their views differ from those who favor " VIVISECTION WITHOUT RESTRICTIONS " of any kind. It would, of course, be a mistake to count these names with those of the majority who favor the leading principles of the statement and restricted vivisection.]

1. Signed as written, without Change.

HERBERT SPENCER, Author, London.

Sir EDWIN ARNOLD, Author, and Editor of the " London Telegraph." London : —

" It is with this that I agree, detesting and dreading unlicensed vivisection. But I love and honour Science too much to deny her any right, exercised with true scientific spirit ; that is, with reverence, mercy, and love to all living things. I would hardly allow even an angel to vivisect without anæsthetics."

The fourth statement, Sir Edwin Arnold characterizes as " the language of scientific devils."

ROBERT BRAITHWAITE, M. D., F. L. S., London : —

" . . . After facts have been sufficiently established, it is not necessary to repeat experiments for individual satisfaction, still less for demonstration to students ; the facts should be accepted from the teacher equally with other facts which cannot be demonstrated."

President DAVID H. COCHRAN, Ph.D., LL.D., Polytechnic Institute, Brooklyn, N. Y.

President MARTIN KELLOGG, LL.D., University of California, Berkeley, Cal.

President HENRY WADE ROGERS, LL.D., Northwestern University, Evanston, Ill.

President ELMER H. CAPEN, D. D., Tufts College, Mass.

President CHARLES KENDALL ADAMS, LL.D., University of Wisconsin, Madison, Wis.

R. H. THURSTON, LL.D., Director of Sibley College, Cornell University, Ithaca, N. Y.

C. C. EVERETT, D. D., Dean of Harvard Divinity School, Cambridge, Mass.

GEORGE HODGES, D. D., Dean of Episcopal Theological School, Cambridge, Mass.

JAMES O. MURRAY, Dean of Princeton University, Princeton, N. J.

CYRUS NORTHROP, LL.D., President University of Minnesota, Minneapolis, Minn. : —

> "Vivisection is practised more than is necessary; it ought undoubtedly to be restrained. Doubtless it has its uses in teaching, but its value in investigation has been overrated."

Prof. WILLIAM JAMES, M. D., Author; Professor of Psychology, Harvard University, Cambridge, Mass. : —

> "If public opinion could constitute the check, I should prefer that; but that would necessarily be ineffectual. I think there will be *great* difficulty in defining by law what is legitimate, or in having whatever law were made, discriminatingly administered. In *principle*, however, I have not a moment's hesitation in standing up for the vivisector being outwardly responsible for his acts."

Rt. Rev. HENRY A. NEELY, Bishop of Maine : —

> "The above statement most nearly expresses my views. If the opinions of Sir Charles Bell, Dr. Lawson Tait, and Dr. Bell-Taylor (quoted in the first statement) were generally endorsed by pathologists of the highest class, it would follow, *me judice*, that vivisection can in no case be justified."

Rt. Rev. JOSEPH BLOUNT CHESHIRE, Jr., Bishop of South Carolina :

> "While as a matter of sentiment I am strongly inclined to say that vivisection should be absolutely prohibited, yet I am not able to justify that position fully to my mind and conscience. The statement of 'vivisection restricted by utility,' seems to me to be in accordance with relations which God has established and declared between man and the lower orders of living creatures. Vivisection should be allowed only in case of necessity or of great utility, and then under strict regulations."

Rt. Rev. MAHLON N. GILBERT, Asst. Bishop of Minnesota.

Rt. Rev. CHARLES TODD QUINTARD, D. D., Bishop of Tennessee.

Rt. Rev. ANSON R. GRAVES, LL.D., Bishop of The Platte.

Rt. Rev. ALEXANDER BURGESS, Bishop of Quincy.

Rt. Rev. CORTLANDT WHITEHEAD, Bishop of Pittsburg.

Rt. Rev. GEORGE W. PETERKIN, LL.D., Bishop of West Virginia.

Rt. Rev. F. D. HUNTINGTON, Bishop of Central New York.

Rt. Rev. LEMUEL H. WELLS, Bishop of Spokane.

Rt. Rev. NELSON SOMERVILLE RULISON, Assistant Bishop of Central Pennsylvania.

Prof. T. M. BALLIET, M. D., Professor of Therapeutics, Dartmouth Medical College, Philadelphia, Pa.

Prof. T. GAILLARD THOMAS, M. D., College of Physicians and Surgeons, Consulting Surgeon of the State Women's Hospital, N. Y.

SIMON BARUCH, M. D., Physician to the Manhattan General Hospital, N. Y., and late Physician and Surgeon to the N. Y. Juv. Asylum.

(Would permit vivisection for demonstration, under anæsthesia.)

Prof. GEO. MONTGOMERY TUTTLE, M. D., Professor of Gynæcology in the College of Physicians and Surgeons, New York.

Prof. ANDREW H. SMITH, M. D., Post-Graduate School, Attending Physician of the Presbyterian Hospital, New York.

Prof. ALONZO BOOTHBY, M. D., Associate Professor of Surgery, Boston, School of Medicine : —

"It does not seem clear upon what grounds you are making the inquiry ; but as the matter is a very important one, and as there has been such an unnecessary and absurd use of animals to amuse students and idlers, I send you my protest, with the hope that your object is to lessen the evil."

DANIEL COOK, M. D., New York City : —

"In my experience, certain vivisections are performed mostly for the most unworthy object of making the lectures sensational above those at other colleges, — exactly as our theatres and newspapers vie with one another in furnishing blood-curdling plays, or sensational news."

JOHN ALLAN WYETH, M. D., President of the Faculty of the New York Polyclinic Medical School and Hospital, New York.

ALBERT McSCULLY, M. D., M. Ch., L. M., Queen's University, formerly Assistant Demonstrator of Anatomy in Queen's College, Ireland, New York City : —

"A person actively engaged in vivisection is inclined to subscribe to the fourth statement. The whole mind is absorbed in the subject ; and clear unbiassed reasoning is then out of the question. I felt thus myself, at one time, when full of my subject, as well as full of youthful ardor. After mature deliberation, I freely and unconditionally subscribe to this statement."

G. B. HOPE, M. D., New York City : —

"From what I have been witness to in several steps of my student career, I am heartily in sympathy with your investigation. I believe in the severest control governing vivisection. Every class-room exhibition particularly should be prohibited as useless and demoralizing. I would have every experimenter file an application, giving the nature and intention of the operation, and subsequently report the number, size, and quality of animals employed, with the results obtained. Such a course would check needless and vicious operations."

(Dr. Hope graduated twenty years ago from a medical college notorious for its extreme vivisections.)

ARCHIBALD T. BANNING, M. D., Pres. City Medical Association, Mt. Vernon, N. Y. : —

" I well remember when a student the feelings of horror that arose on seeing certain experiments. . . . The first experiment was altogether an outrage ; the second, though of some utility, had already been sufficiently demonstrated, and a mere statement from the professor would have accomplished as much instruction as ocular evidence. The impression thus made on the unformed minds of students is bad, and might have a tendency to develop some morbid psychopathic action such as ' Sadism.' I have such cases in view."

JOHN L. SCHOOLCRAFT, M. D., Schenectady, N. Y. : —

" The continual practice of vivisection by assistant lecturers and others to show what has been thoroughly proven by men of reputation, should be absolutely prohibited."

WILLIAM J. BURR, M. D., late Acting Staff-Surgeon, U. S. A., Newark Valley, N. Y. : —

" I have seen most kindly conducted experiments, and also others most abhorrent. In my opinion, vivisection should be under restrictions, and conducted without pain."

S. P. MOORE, M. D., Munnville, N. Y. : —

" I am aware that we are apt to forget what is right in efforts after fame. As I grow older, certain scenes before a class of young men seem to me of very doubtful propriety. Medical students are apt to be rough enough without such sights."

JONATHAN KNEELAND, M.D., S. Onondaga, N. Y. : —

" If we know less of the mysteries of existence by refraining from tormenting our pets, we shall at any rate increase the total joy of animal life."

WILLIAM H. MUNN, M. D., New York City : —

" No undergraduate to attempt it ; only by a professor, and with the least pain."

JOHN PARMENTER, M. D., Prof. of Anatomy and Clinical Surgery, University of Buffalo, Surgeon to the Erie Co. Fitch and Children's Hospitals, Buffalo, N. Y.

ARCHIBALD M. CAMPBELL, M. D., Consulting Physician in the Home for Incurables, N. Y. City, Member of the Academy of Medicine, Physician to the N. Y. Infant Asylum, etc., Mt. Vernon, N. Y.

HERMAN MYNTER, M. D., Professor of Surgery, Niagara University, Buffalo, N. Y.

JAMES E. KELLY, M. D., F. R. C. S., Consulting Surgeon, French Hospital, New York City.

O. B. DOUGLASS, M. D., Surgeon to the Manhattan Eye and Ear Hospital, late President of the Medical Society of the county of New York, etc.

CHARLES S. MACK, M. D., Professor of Materia Medica and Therapeutics, University of Michigan, Ann Arbor, Mich. : —

" . . . Regard as useless much that some regard as useful."

GEORGE M. GOULD, M. D., Editor of the " Medical News," Philadelphia, Pa. : —

" Whenever possible, under anæsthesia."

(Dr. Gould's views regarding vivisection have been well expressed in his recent work on biological investigation, " The Meaning and Method of Life," from which we quote and italicize the following passage : " If a very limited use of vivisection experiment is necessary for scientific and medical progress, *it must be regulated by law*, carried out with jealous guarding against excess and against suffering, and the maimed animals painlessly killed when the experiment is complete. *The practice carried on by conceited jackanapes to prove over and over again already ascertained results, to minister to egotism, for didactic purposes, — these are not necessary, and must be forbidden.*")

ISAAC SHARPLESS, LL.D., President of Haverford College, Pa.

A. H. FETTEROLF, Ph.D., LL.D., President of Girard College, Philadelphia, Pa.

W. F. McDOWELL, D. D., Chancellor of the University of Denver, Colorado.

WM. M. BLACKBURN, D. D., LL.D., President of Pierre University, Pierre, S. D.

W. H. SCOTT, LL.D., President of the Ohio State University, Columbus, Ohio.

FRANCIS WAYLAND, LL.D., Dean of Yale Law School, New Haven, Ct.

FRANKLIN W. HOOPER, Director of the Institute of Arts and Sciences, Brooklyn, N. Y.

EDWIN J. HOUSTON, Electrical Expert, Philadelphia, Pa.

EDWARD S. HOLDEN, LL.D., Astronomer, Director of Lick Observatory, Mt. Hamilton, California.

Prof. HENRY FAIRFIELD OSBORN, Da Costa Professor of Biology, Columbia Coll., N. Y.

Prof. JAMES L. ROBERTSON, M. D., Professor of Theory and Practice, Am. Veterinary College, N. Y. City.

Rev. SAMUEL A. BARNETT, Warden of Toynbee Hall, London, England.

HAMO THORNYCROFT, R. A., Sculptor, Loudon, England.

Dr. HAVELOCK ELLIS, Editor of the "Contemporary Science Series," Cornwall, England.

EDWARD BRECK, Ph.D., Journalist, Berlin : —

"A limitation to one or two laboratories in each State might be wise."

FRANCIS F. BROWNE, Editor of "The Dial," Chicago, Ill. : —

"Believing that all the relations of men to animals, like the relations of men to each other, should be subject to State regulation, I of course hold that vivisection should be under such control, and very stringently."

Prof. WM. A. PACKARD, Princeton University, New Jersey.

Prof. GEORGE M. HARPER, Ph.D., Princeton University, N. J.

Prof. BENJ. IDE WHEELER, Cornell University, Ithaca, N. Y.

Prof. EDWARD HITCHCOCK, Jr., Cornell University, Ithaca, N. Y.

Prof. CHARLES E. BENNETT, Cornell University, Ithaca, N. Y.

Prof. GEORGE P. BRISTOL, Cornell University, Ithaca, N. Y. : —

"In hearty sympathy with this statement."

Prof. ROBERT BAIRD, Northwestern University, Ill.

Prof. CHARLES F. BRADLEY, D.D., Garrett Biblical Institution, Evanston, Ill.

Prof. JOHN M. SHALLER, M.D., Professor of Physiology, College of Medicine and Surgery, Cincinnati, Ohio.

Prof. C. F. BRACKETT, M.D., LL.D., President of the Board of Health for the State of New Jersey, Professor of Physics, Princeton, N. J.

Prof. WM. FRANCIS MAGIE, Ph.D. (Berlin), Professor of Physics, Princeton College, N. J.

Prof. FRANCIS H. HERRICK, Biologist, Adelbert College, Cleveland, O.

Prof. OGDEN N. ROOD, Professor of Physics, Columbia College, New York.

Prof. CHARLES B. ATWELL, Ph.M., Professor of Botany, N. W. University, Evanston, Ill.

Prof. A. E. TURNER, A.M., Professor Natural Sciences, Lincoln University, Ill.

Prof. HENRY B. CORNWALL, Professor of Chemistry, Princeton University, N. J.

Prof. HENRY L. OBETZ, M.D., Professor of Surgery, University of Michigan, Ann Arbor.

Prof. FREDERICK TRACY, Ph.D., Lecturer in Psychology, University of Toronto, Canada.

Prof. JOHN C. BRANNER, Ph.D., Professor of Geology, Stanford University, Cal.

Prof. ALBERT NOTT, M. D., Professor of Physiology and Dean of Tufts College Medical School, Mass.

Prof. ALBERT E. MILLER, M. D., Professor of Physiology, College of Physicians and Surgeons, Boston, Mass.

Prof. A. A. D'ANCONA, M. D., Professor of Physiology, University of California, San Francisco, Cal.

Pres. ALVAH HOVEY, D. D., LL D., President of the Newton Theological Institute, Newton, Mass.

Prof. ARTHUR S. HOYT, A. M., D. D., Auburn Theological Seminary, Auburn, N. Y.

Prof. GEO. R. FREEMAN, A. M., D. B., Meadville Theological School, Penn.

Prof. WOOSTER W. BEMAN. A. M., University of Michigan, Ann Arbor, Mich.

Prof. BENJ. S. TORREY, University of Chicago, Ill.

Prof. W. H. MACE, A. M., Syracuse University, N. Y.

Prof. FRANKLIN J. HOLZWARTH, Ph.D., Syracuse University, N. Y.

Prof. EUGENE HAANEL, Ph.D., Professor of Physics, Syracuse University, N. Y.

Prof. JOHN R. FRENCH, Vice-Chancellor, Syracuse University, N. Y.

Prof. L. A. SHERMAN, Lincoln, Neb.

Prof. A. T. MURRAY, Stanford University, Cal.

Prof. LEVERITT W. SPRING, Williams College, Mass.

Prof. L. D. WOODBRIDGE, M. D., Williams College, Mass.

Prof. ANSON D. MORSE, M. A., Amherst College, Mass.

Prof. CLIFFORD H. MOORE, University of Chicago, Ill.

Prof. JOHN B. CLARK, Amherst College, Mass.

Prof. J. B. PARKINSON, University of Wisconsin, Madison, Wis.

Prof. WALTER D. TOY, University of North Carolina.

Prof. E. W. HYDE, University of Cincinnati, Ohio.

Prof. CHARLES E. FAY, A. M., Tufts College, Mass.

Prof. EDWARD A. ALLEN, Lit. Dr., University of Missouri, Columbia.

Prof. ISAAC N. DEMMON, University of Michigan, Ann Arbor.

Prof. CHARLES DAVIDSON, Ph.D., Adelbert College, Cleveland, O.

Prof. G. T. KNIGHT, Tufts College, Mass.

Prof. CORNELIUS B. BRADLEY, A. M., University of California, Berkeley, Cal. : —

"I should wish to go a step further, and while not *absolutely* limiting vivisection to painless forms, as is done in statement II., I think that the manner of it should always be determined by considerations of humanity; it should be painless *in all cases where painlessness will serve the purposes* of an otherwise useful and desirable investigation. While I am aware of the difficulty of enforcing this distinction, I think it important not to attempt to make pain the *sole* factor in the decision."

Prof. Arthur T. Hadley, Yale University, New Haven.

Prof. Frederic D. Allen, Ph.D., Harvard University, Cambridge.

Prof. Eugene L. Richards, Yale University, New Haven.

Prof. C. G. Rockwood, Jr., Princeton University, New Jersey.

Francis A. Schlitz, M. D., Brooklyn, N. Y.

D. H. Goodwillie, M. D., New York.

Thos. Gilfillan, M. D., Northampton, Mass.

Prof. James H. Robinson, Ph.D., Columbia College, New York.

Prof. J. Macy, A. M., Iowa College, Grinnell, Iowa.

Allen M. Thomas, M. D., President of the N. Y. Clinical Society, New York City.

J. Oscoe Chase, M. D., Assistant Surgeon of the N. Y. Opthalmic Hospital, etc., New York City.

John L. Hildreth, M. D., Cambridge, Mass.

Prof. J. S. Prout, M. D., Long Island College Hospital, Brooklyn, N. Y.

D. Branch Clark, M. D., New York City.

Abel Huntingdon, M. D., New York City.

Jared G. Baldwin, M. D., New York City:

"While I favor vivisection as necessary at times for the progress of science, and would restrict it by utility, I realize the great difficulty there will be in drawing the line between utility and uselessness. . . . Still, I think some restrictions should be made."

Geo. P. Shirmer, M. D., New York City:

"I believe vivisection is also justifiable for the purpose of teaching facts already known, when no pain is caused either as an immediate or subsequent result of the operation."

Gorham Bacon, M. D., New York City.

Richard T. Bang, M. D., New York City.

Rollin B. Gray, M. D., New York City.

Frederick Guttman, M. D., New York City.

Joseph T. O'Connor, M. D., New York City.

Joseph Eichberg, M. D., New York City.

H. O. Clauss, M. D., New York City.

Joseph Braunstein, M. D., New York City.

E. P. Miller, M. D., New York City.

Ervin A. Tucker, M. D., New York City.

T. C. Williams, M. D., New York City.

Alexander Halden, M. D., New York City.

Andrew F. Currier. M. D., New York City:

"This represents, in the main, my views on the subject."

H. S. Drayton, M. D., New York City.

Geo. L. Simpson, M. D., New York City.

Thos. Wilde, M. D., New York City.

Frank A. McGuire, M. D., New York City.

G. H. Patchen, M. D., New York City.

E. D. Franklin, M. D., New York City.

L. L. Bradshaw, M. D., New York City.

John G. Perry, M. D., New York City.

S. Wesley Smith, M. D., New York City.

Ira B. Read, M. D., New York City.

Caroline L. Black, M. D., New York City.

R. C. M. Page, M. D. (Prof. New York Polyclinic), New York City.

J. Henry Fruitnight, M. D., New York City.

C. Ruxton Ellison, M. D., New York City:

"This statement agrees with my ideas of the subject precisely."

Garret Cosine, M. D., New York City.

E. B. Foote, Jr., M. D., New York City.

Stephen J. Clark, M. D., New York City.

A. H. Heath, M. D., New York City.

Sanford J. Murray, M. D., New York City.

Harriette C. Keating, M. D., Sc. D., New York City.

Floyd T. Sheldon, M. D., New York City.

Seth D. Close, M. D., New York City.

James A. Bennett, M. D., New York City:

"Allowable for demonstration, if without pain."

Octavius A. White, M. D., New York City.

(Adds last clause of statement III.)

Joseph Kucher, M. D., New York City.

Frank W. Merriam, M. D., New York City.

H. M. Hitchcock, M. D., New York City.

Paul C. Bonmer, M. D., Prof. of Anatomy, etc., Chicago, Ill.

John J. Orton, M. D., Lakeville, Conn.

J. N. Martin, M. D., Ann Arbor, Mich.

D. M. Cattell, M. D., Chicago, Ill.

Mary H. Thompson; M. D., Chicago, Ill.
Frank Billings, M. D., Chicago, Ill.
J. B. Murdock, M. D., Pittsburgh, Pa.
Louis T. Riesmeyer, M. D., St. Louis, Mo.
Francis B. Hill, M. D., Colorado Springs, Col.
Wm. S. Stewart, M. D., Philadelphia, Pa.
(Signs III. also.)
J. D. Blake, M. D., Baltimore, Md.
Charles G. Hill, M. D., Baltimore, Md.
K. W. Baldwin, M. D., Philadelphia, Pa.
L. S. Kelsey, M. D., Richmond, Ind.
D'Estaing Dickerson, M. D., Kansas City, Mo.
George W. Cale, M. D., St. Louis, Mo.
Edwin R. Maxson, M. D.
Aven Nelson, M. D., A. M., Prof. of Biology
in the University of Wyoming, Laramie:
" All experiments should be painless, so far
as possible; and so far as it will not
interfere with success of the experi-
ments." (The last clause nullifies the
statement.)
George E. Paul, M. D., Rutland, Vt.
Ross Wilson, M. D., Chicago, Ill.
H. S. Maxson, M. D., St. Helena, Cal.
W. H. Maxson, M. D., St. Helena, Cal.:
" As far as possible without pain."
Edmund J. A. Rogers, M. D., Denver, Col.
James W. Ovenpeck, M. D., Hamilton, Ohio.
Ida B. Hunt, M. D., Plainfield, N. J.
L. M. Giffin, M. D., Boulder, Col.
W. A. Lockwood, M. D., Norwalk, Conn.
Prof. Walter S. Haines, M. D., Rush Medical
College, Chicago.
Prof. J. M. Withrow, M. D., Cincinnati, Ohio.
John H. Thompson, M. D., New York City:
" Allowable only under an anæsthetic."
Louis N. Schultz, M. D., New York City.
Edwin West, M. D., New York City.
J. A. Towner, M. D., New York City.
Samuel G. Sewall, M. D., New York City.
Henry Tuck, M. D., New York City.
William F. Wright, M. D., New York City.
Thomas F. Smith, M. D., New York City.
Frank Livermore, M. D., New York City.
H H. Kane, M. D., New York City.
O. S. Phelps, M. D., New York City.
Homer I. Ostrom, M. D., New York City.
P. J. Lynch, M. D., New York City.
Granville C. Brown, M. D., New York City.
Adoniram B. Judson, M. D., New York City.
Reuben B. Burton, M. D., New York City.
Egbert Guernsey, M. D., New York City.
John E. Comfort, M. D., New York City.
Charles Milne, M. D., New York City.
John P. Nolan, M. D., New York City.
George B. Durrie, M. D., New York City.

J. E. Janvrin, M. D., New York City.
Richard E. Kunze, M. D., New York City.
Ephraim Cutter, M. D., LL.D., New York
City. (See extract from letter, p. 65.)
George E. Tytler, M. D., New York City:
" Persons sentenced to death would serve a
most useful purpose if, before execution
of the sentence, they were subjected to
experimentation in testing new reme-
dies, etc."
William L. Flemming, M. D., New York City.
Henry E. Crampton, M. D., New York City.
W. P. Northrup, M. D., New York City.
Willard Parker, M. D., New York City.
Prof. William E. Rounds, M. D., Professor in
the N. Y. Opth. Hospital College, New
York City.
Emily Blackwell, M. D., Dean of the Woman's
Medical College, New York City.
(Agrees with last clause of No. III.)
S. A. Russell, M. D., Poughkeepsie, N. Y.:
" And as far as possible painless."
F. A. Winne, M. D., Brockport, N. Y.
John B. Ellis, M. D., Little Falls, N. Y.
J. H. Trumbull, M. D., Hornellsville, N. Y.
Chas. P. Russell, M. D., Utica, N. Y.
Chas. S. Starr, M. D., Rochester, N. Y.
Thos. M. Flandrau, M. D., Rome, N. Y.:
" Should be invariably mitigated by anæs-
' thetics."
Geo. H. Noble, M. D., Cairo, N. Y.
Geo. M. Palmer, M. D., Warsaw, N. Y.
Pascal M. Dowd, M. D., Oswego, N. Y.
Geo. B. Chapman, M. D., Dover Plains, N. Y.
M. M. Bagg, M. D., Utica, N. Y.
S. R. Welles, M. D., Waterloo, N. Y.
C. D. Spencer, M. D., Binghamton, N. Y.:
" I most heartily endorse the above senti-
ments."
A. Miller, M. D., Jordanville, N. Y.
Arthur R. Hill, M. D., Farmer, N. Y.
E. G. Williams, M. D., Remsen, N. Y.
J. W. Douglass, M. D., Boonville, N. Y.
H. Sheldon Edson, M. D., Cortland, N. Y.
A. T. Van Vranken, M. D., West Troy, N. Y.
(With anæsthetics, in every instance
possible.)
Henry F. Kingsley, M. D., Schoharie, N. Y.
John C. Fisher, M. D., Warsaw, N. Y.
Arthur E. Tuck, M. D., Gloversville, N. Y.
E. W. Gallup, M. D., Stamford, N. Y.
C. C. Thayer, M. D., Clifton Springs, N. Y.
J. E. Smith, M. D., Clyde, N. Y.
Wm. H. Hodgman, M. D., Saratoga, N. Y.
Paris G. Clark, M. D., Unadilla, N. Y.

T. D. Spencer, M. D., Rochester, N. Y.

William E. Hathaway, M. D., Hornellsville, N. Y.

J. D. Mitchell, M. D., Hornellsville, N. Y.

Thomas B. Fowler, M. D., Cohocton, N. Y.

O. S. Martin, M. D., Salamanca, N. Y.

E. H. Loughran, M. D., Kingston, N. Y.

J. L. Gardiner, M. D., Bridgehampton, N. Y.

W. Scott Hicks, M. D., Bristol, N. Y.

A. S. Zabriskie, M. D., Suffern, N. Y.

W. B. Putnam, M. D., Hoosick Falls, N. Y.

Charles B. Hawley, M. D., Gouverneur, N. Y.

J. H. Weckel, M. D., Breakabeen, N. Y.

Samuel J. Crockett, M.D., Sandy Creek, N. Y.

T. Millspaugh, M. D., Wallkill, N. Y.

Charles E. Witbeck, M. D., Cohoes, N. Y.

Dr. Edwin R. Maxson, LL.D., Syracuse, N. Y.

H. A. Place, M. D., Ceres, N. Y.

James H. Glass, M. D., Surgeon-in-charge, Utica City Hospital, Ex-President County Medical Society, etc., Utica, N. Y. : —

"The full significance of the last sentence cannot be too strongly emphasized."

Charles G. Stratton, M. D., Buffalo, N. Y.

E. Rainier, M. D., Oswego, N. Y.

S. Wright Hurd, M. D., Lockport, N. Y.

John J. Montgomery, M. D., Dryden, N. Y.

M. B. Folwell, A. M., M. D., Clinical Professor of Diseases of Children, University of Buffalo, N. Y.

George H. Noble, M. D., Cairo, N. Y.

S. P. Welles, M. D., Waterloo, N. Y.

Ransom Terry, M. D., Ischua, N. Y.

Guy R. Cook, M. D., Syracuse, N. Y.

James Allen, M. D., Richford, N. Y.

Willard R. Fitch, M. D., Knowlesville, N. Y.

F. A. Dutton, M. D., Gainesville, N. Y.

(Adds the last clause of statement III.)

G. W. Murdock, M. D., Cold Spring, N. Y.

Charles Forbs, M. D., Rochester, N. Y.

G. W. Faller, M. D., Oyster Bay, N. Y.

R. N. Cooley, M. D., Hannibal Centre, N. Y.

J. D. Guy, M. D., Chenango Forks, N. Y.

William B. Mann, M. D., Brockport, N. Y.

(Erases " in private.")

R. H. Morey, M. D., Old Chatham, N. Y.

W. M. Hilton, M. D., Waverly, N. Y.

Fred A. Wright, M. D., Glen Cove, N. Y.

R. S. Prentiss, M. D., Long Island City, N.Y.

Donald McPherson, M. D., Palmyra, N. Y.

A. J. Mixsell, M. D., Mamaroneck, N. Y.

J. W. Huntington, M. D., Mexico, N. Y.

W. C. Earl, M. D., Buffalo, N. Y.

O. J. Hallenbeck, M. D., President Ontario Co. Medical Society, Canandaigua, N. Y.

M. W. Vandenburg, A. M., M. D., Fort Edward, N. Y.

H. T. Dana, M. D., Cortland, N. Y.

F. A. Strong, M. D., Brewerton, N. Y.

Emily H. Wells, M. D., Binghamton, N. Y.

John E. Weaver, M. D., Rochester, N. Y.

Mrs. M. L. D. Wilson, M. D., Troy, N. Y.

A. P. Carsons, M. D., Forestville, N. Y.

George V. R. Merrill, M. D., Elmira, N. Y.

M. R. Carson, M. D., Canandaigua, N. Y.

J. D. Featherstonhaugh, M. D., Cohoes, N. Y.

George M. Abbott, M. D., Castleton, N. Y.

Wallace Sibley, M. D., Rochester, N. Y.

James M. Barrett, M. D., Owego, N. Y.

G. W. Seymour, M. D., Westfield, N. Y.

Theo. Waiser, M. D., New Brighton, N. Y.

Smith Ely, M. D., Newburgh, N. Y.

Elizabeth R. G. Myer, M. D., Turner, N. Y.

Nathan P. Tyler, M. D., New Rochelle, N .Y.

Marcenos H. Cole, M. D., Newfane, N. Y.

E. B. Tefft, M. D., New Rochelle, N. Y.

A. M. Comfort, M. D., Syracuse, N. Y.

Byron Pierce, M. D., Coopers Plains, N. Y.

De Witt C. Jayne, M. D., Florida, N. Y.

M. M. Frye, M. D., Auburn, N. Y.

E. T. Rulison, M. D., Amsterdam, N. Y.

J. W. Gee, M. D., Van Etten, N. Y.

Carlos T. Miller, M. D., Mount Kisco, N. Y.

S. H. Freeman, M. D., Albany, N. Y.

Thomas Becket, M. D., Albany, N. Y.:

" Without pain."

Edward Torrey, M. D., Allegany, N. Y.

H. E. Allison, M. D., Medical Supt. Matteawan State Hospital, Fishkill, N. Y.

Henry C. Coon, M. D., Alfred, N. Y.

T. Kirkland Perry, M. D., Albany, N. Y.

L. E. Rockwell, M. D., Amenia, N. Y.

J. R. Fairbanks, M. D., Amsterdam, N. Y.

P. J. Keegan, M. D., Albany, N. Y.

Frederic C. Curtis, M. D., Albany, N. Y.

Richard M. Moore, M. D., Rochester, N. Y.

William S. Cheesman, M. D., Auburn, N. Y.

W. H. Procter, M. D., Binghamton, N. Y.

Clayton M. Daniels, M. D., Buffalo, N. Y.

Stephen Y. Howell, M. D. (M. R. C. S., England), Buffalo, N. Y.

C. E. Heaton, M. D., Baldwinsville, N. Y.

W. J. Nellis, M. D., Albany, N. Y.

B. H. Grove, M. D., Buffalo, N. Y.

Wm. C. Phelps, M. D., Buffalo, N. Y.

F. Findlay, M. D., Franklinville, N. Y.

James L. Cooley, M. D., Glen Cove, N. Y.

C. H. Masten, M. D., Sparkill, N. Y.

J. Harris Oxner, M. D., Rome, N. Y.

D. D. Drake, M. D., Johnstown, N. Y.

Thomas B. Nichols, M. D., Plattsburg, N. Y.

Charles R. Weed, M. D., Utica, N. Y.

Charles H. Langdon, M. D., Phys. to Hudson River State Hospital, Poughkeepsie, N.Y.

M. L. Chambers, M. D., Port Jefferson, N. Y.

Alva D. Decker, M. D., Prince Bay, N. Y.

Paul D. Carpenter, M. D., Pittsford, N. Y.

E. W. Capron, M. D., Lansingburgh, N. Y.

Orson G. Dibble, M. D, Pompey, N. Y.

H. C. Hendrick, M. D., McGrawville, N. Y.

Arthur B. Kinne, M. D., Syracuse, N. Y.

Edwin Barnes, M. D., Pleasant Plains, N. Y.

E. M. Lyon, M. D., Plattsburg, N. Y.

John C. DuBois, M. D., Hudson, N. Y.:
"If possible, without causing pain."

Valentine Browne, M. D., Yonkers, N. Y.

D. V. Still, M. D., Johnstown, N. Y.

T. H. Cox, M. D., Lee Centre, N. Y.

Lucius B. Parmele, M. D., Batavia, N. Y.

Lyman Barton, M. D., Willsboro, N. Y.

E. Howe Davis, M. D., Elmira, N. Y.

Cordelia A. Greene, M. D., Castile, N. Y.

A. C. Grover, M. D., Port Henry, N. Y.

Albert W. Palmer, M. D., Marlborough, N. Y.

Randall Williams, M. D., Ex-President of the Genesee Co. Med. Society, Le Roy, N. Y.

F. G. Osborne, M. D., South Wales. N. Y.

S. Walter Scott, M. D., Troy. N. Y.

F. P. Beard, M. D., Cobleskill, N. Y.

J. K. Stockwell, M. D., Oswego, N. Y.

E. A. Chapman, M. D., Belleville, N. Y.:
"Without pain, when possible."

J. Erwin Reed, M. D., Carmel, N. Y.

Henry Sperbeck, M. D., Charlotteville, N. Y.

E. W. Earle, M. D., Rochester, N. Y.

O. W. Peck, M. D., Oneonta, N. Y.

Seldon J. Mudge, M. D., Olean, N. Y.

J. G. Russell, M. D., Salem, N. Y.

Osman F. Kinlock, M. D., Troy. N. Y.

J. Seward White, M. D., Glen Falls, N. Y.

Charles G. Bacon, M. D., Fulton, N. Y.

John J. Walsh, M. D., Buffalo, N. Y.

D. A. Barnum, M. D., Cassville, N. Y.

I. G. Johnson, M. D., Greenfield Centre, N. Y.

J V. D. Coon, M. D., Olean, N. Y.

J. H. Wiggins, M. D., Jamestown, N. Y.

Newton F. Curtis, M. D., White Plains, N. Y.

Fremont W. Scott, M. D., Medina, N. Y.

Jerome H. Coe, M. D., Syracuse, N. Y.

L. T. White, M. D., Homer, N. Y.

Geo. Huntington, M. D., La Grangville, N. Y.

H. P. Whitford, M. D., Bridgewater, N. Y.

M. H. Bronson, M. D., Lowville, N. Y.

H. D. Weyburn, M. D., Geneva, N. Y.

S. J. Pearsall, M. D., Saratoga Springs, N. Y.

John C. Sill, M. D., Argyle, N. Y.

Jacob L. Williams, M. D., Boston, Mass.
(Signs both II. and III.)

Eli H. Long, M. D., Professor of Materia Medica in Buffalo College of Pharmacy, Buffalo, N. Y.

Francis Brick, M. D., Vice-Pres. Mass. Surg. and Gynæ. Soc., Worcester, Mass.

O. J. Brown, M. D., North Adams, Mass.

F. W. Brigham, M. D., Shrewsbury, Mass.

John D. Young, M. D., Winthrop, Mass.

G. E. Fuller, M. D., Monson, Mass.

J. Anson Bushee, M. D., East Boston, Mass.

Locero J. Gibbs, M. D., Chicopee, Mass.

W. P. M. Ames, M. D., Springfield, Mass.

Stephen W. Driver, M. D., Cambridge, Mass.

James Peirce. M. D., Methuen, Mass.

John Dixwell, M. D., Boston, Mass.
(Questions the existence of any extensive abuse of vivisection in the United States.)

J. A. Houston, M. D., Northampton. Mass.

Hosea M. Quinby, M. D., Worcester, Mass.

Frank A. Hubbard, M. D., Taunton, Mass.:
"Painlessly, if possible."

Julius Garst, M. D., Worcester, Mass.

L. J. Putnam, M. D., No. Adams, Mass.

Charles W. Haddock, M. D., Beverly, Mass.

C. A. Wheeler, M. D., Leominster. Mass.

Morgan L. Woodruff, M. D., Pittsfield, Mass.

Franz H. Krebs, M. D., Boston, Mass.: —
"The practice should only be allowed to Zoölogists of the first rank."

Fred. W. Chapin, M. D., Springfield, Mass.

O. W. Phelps, M. D., Warren, Mass.

T. Haven Denring. M. D., Braintree, Mass.

Alfred A. Mackeen. M. D., Whitman, Mass.

George C. Osgood, M. D., Lowell, Mass.

Edward H. Adams, M. D., Plymouth, Mass.:
"Vivisection should also be restricted by regard to the highest dictates of humanity."

John A. Gordon. M. D., Quincy, Mass.

F. A. Rogers, M. D., Chatham, Mass.

S. W. Bowen, M. D., Fall River, Mass.

Geo. F. Simpson, M. D., North Adams, Mass.
(Adds the last three lines of Statement III.)

Benj. M. Burrell. M. D., Boston, Mass.: —
"An anæsthetic should always be used, if possible."

E. A. Daniels, M.D. (Harvard), Boston, Mass.

S. T. Hyde, M. D., Dorchester, Mass.

James L. Harriman, M. D., Hudson, Mass.

Thomas Waterman, M. D., Boston, Mass.

Clarence L. Hower, M. D., Hanover, Mass.

Frederick F. Doggett, M. D. (Harvard), Boston, Mass.

Dwight E. Cone, M. D., Fall River, Mass.

Emma L. Call, M. D., Boston, Mass.
Rufus K. Noyes, M. D., Boston, Mass.
F. Gordon Morrill, M. D., Boston, Mass.
Edward C. Briggs, M. D. (Assistant Professor of Materia Medica, Harvard University), Boston, Mass.
E. S. Boland, M. D., So. Boston, Mass.
Henry N. Jones, M. D., Kingston, Mass.
W. H. Sylvester, M. D., Natick, Mass.
Henry M. Chase, M. D., Lawrence, Mass.
Helen A. Goodspeed, M. D., Leicester, Mass.
John H. Gilbert, M. D., Quincy, Mass.
James A. Dow, M. D., Cambridge, Mass.
A. S. Osborne, M. D., Medford, Mass.
George E. Percy, M. D., Salem, Mass.
Wm. K. Knowles, M. D., Everett, Mass.:
  "This statement corresponds with my views. Vivisection should certainly be carefully restricted."
C. N. Chamberlain, M. D., Andover, Mass.:
  "Anæsthesia should be employed in every case where its use would not defeat the object of inquiry."
S. L. Eaton, M. D., Newton Highlands, Mass.
J. J. B. Vermyne, M. D., New Bedford, Mass.
Charles Jordan, M. D., Wakefield, Mass.
Lincoln R. Stone, M. D., Newton, Mass.
J. Winthrop Spooner, M. D., Hingham, Mass.
John M. French, M. D., Milford, Mass.
Rollin C. Ward, M. D., Northfield, Mass.
Willard S. Everett, M. D., Hyde Park, Mass.
Otis H. Johnson, M. D., Haverhill, Mass.
H. A. Smith, M. D., Bondsville, Mass.
Edwin B. Harvey, M. D., Westborough, Mass.
J. H. Robbins, M. D., Hingham, Mass.
(Signs also No. III.)
F. H. Davenport, M. D., Instructor in Gynæcology, Harvard Medical School, Boston, Mass.
A. A. Arthur, M. D., Marshfield, Mass.
J. A. Follett, M. D., Boston, Mass.
John A. Lamson, M. D., Boston, Mass.
James Dunlap, M. D., Northampton, Mass.
Julia A. Marshall, M. D., Haverhill, Mass.
E. E. Spencer, M. D., Cambridgeport, Mass.
Chas. W. Stevens, M. D., Charlestown, Mass.
Nathan French, M. D., Malden, Mass.
Leslie A. Phillips, M. D., Berkeley St., Boston, Mass.
S. F. Haskins, M. D., Orange, Mass.: —
  "Would have an anæsthetic used whenever practical."
Luther G. Chandler, M. D., Townsend, Mass.
J. K. Warren, M. D., Worcester, Mass.
F. J. Canedy, M. D., Shelburne Falls, Mass.
E. Proctor Peirce, M. D., Springfield, Mass.
H. A. Deane, M. D., Easthampton, Mass.

C. M. Barton, M. D., Hatfield, Mass.
  (Adds, "and permitted only to competent and trustworthy persons.")
Edwin A. Colby, M. D., Gardner, Mass.
W. H. Hildreth, M. D., Newton Upper Falls, Mass.
M. V. Pierce, M. D., Milton, Mass.
C. A. Allen, M. D., Holyoke, Mass.: —
  "And without pain."
John B. Learned, M. D., Florence, Mass.:
  "And without pain."
C. C. Messer, M. D., Turners Falls, Mass.
C. G. Trow, M. D., Sunderland, Mass.
C. Blodgett, M. D., Holyoke, Mass.: —
  "And without pain."
C. M. Wilson, M. D., Shelburne Falls, Mass.
John P. Brown, M. D., Taunton, Mass.
Porter Hall, M. D., Leominster, Mass.
John C. Irish, M. D., Lowell, Mass.
Thomas Conant, M. D., Gloucester, Mass.
Moses W. Kidder, M. D., Lincoln, Mass.
John W. Crawford, M. D., Lawrence, Mass.:
  "I would restrict vivisection to medical schools, and to such professors of anatomy and physiology (or medical students under their guidance) as may receive license under State supervision."
P. Wadsworth, M. D., Malden, Mass. : —
  "Painless, so far as possible."
Seraph Frissell, M. D., Springfield, Mass.
Henry J. Kenyon, M. D., Worcester, Mass.:
  "To all the above I most heartily and earnestly subscribe."
Wm. A. McDonald, M. D., Lynn, Mass.
Daniel C. Rose, M. D., Stoughton, Mass.
J. F. Adams, M. D., Worcester, Mass.
A. Carter Webber, M. D., Cambridge, Mass.
S. A. Sylvester, M. D., Newton Centre, Mass.
Orin Warren, M. D., West Newbury, Mass.
W. H. Tobey, M. D., Boston, Mass.
John H. Kennealy, M. D., Boston, Mass.
  (Agrees with last clause of III.)
Henry G. Preston, M. D., Brooklyn, N. Y.
J. Lester Keep, M. D., Brooklyn, N. Y.
Francis H. Miller, M. D., Physician to St. Malachy's Home, Brooklyn, N. Y.
John H. Trent, M. D., Brooklyn, N. Y.
John H. French, M. D., Brooklyn, N. Y.
Arthur Beach, M. D., Brooklyn, N. Y.
Frank Bond, M. D., Brooklyn, N. Y.
J. Freeman Atwood, M. D., Brooklyn, N. Y.
Benjamin Ayres, M. D., Brooklyn, N. Y.
A. Nelson Bell, M. D., Editor of "The Sanitarian," Brooklyn, N. Y.
Wesley Sherman, M. D., Brooklyn, N. Y.

J. L. Cardozo, M. D., D. D., Brooklyn, N. Y.:
"As far as possible, without causing pain."

George Nichols, M. D., Brooklyn, N. Y.
George W. Cushing, M. D., Brooklyn, N. Y.
Jesse B. Lung, M. D., Brooklyn, N. Y.
Henry F. Risch, M. D., Brooklyn, N. Y.
L. H. Miller, M. D., A. M., Brooklyn, N. Y.:
"Most nearly represents my views."

S. E. Stiles, M. D., Brooklyn, N. Y.
Glentworth R. Butler, A. M., M. D., Physician
to the M. E. Hospital, Brooklyn, N. Y.
(Favors vivisection restricted only by
"possible, probable, or certain utility.")

Eliza W. Mosher, M. D., Brooklyn, N. Y.
John F. Davis, M. D., Brooklyn, N. Y.
J. B. Mattison, M. D., Brooklyn, N. Y.
W. Armstrong Fries, M. D., Brooklyn, N. Y.
T. C. Giroux, M. D., Brooklyn, N. Y.
J. G. Atkinson, M. D., Brooklyn, N. Y.
N. A. Robbins, M. D., Surgeon of the Brook-
lyn Fire Department, Brooklyn, N. Y.:
"Vivisection only allowable if without
pain."

Benjamin Edson, M. D., Brooklyn, N. Y.:
"With proper use of anæsthetics."

G. Leroy Menzie, M. D., Oneida, N. Y.
Howell White, M. D., Fishkill, N. Y.: —
"And without pain."

W. P. Clothier, M. D., Buffalo, N. Y.
Geo. F. Perry, M. D., Woodbourne, N. Y.
Mary Armstrong, M. D., Jamestown, N. Y.
W. S. Webster, M. D., Liberty, N. Y.
R. J. Carroll, M. D., Red Hook, N. Y.
Edw. E. Brown, M. D., Glenville, N. Y.
C. R. Rogers, M. D., Newark Valley, N. Y.
James W. Putnam, M. D., Lyons, N. Y.
Amelia E. DeNott, M. D., Syracuse, N. Y.
W. F. Nutten, M. D., Newark, N. Y.
Clifford Hewitt, M. D., Hoosick Falls, N. Y.
W. E. Whitford., M. D., Oxbow, N. Y.
M. M. Fenner, M. D., Fredonia, N. Y.: —
"Use of anæsthetics, when practicable,
should be enjoined.'

Porter Farley, M. D., Rochester, N. Y.
B. B. Bontecou, M. D., Troy, N. Y.
E. V. Denell, M. D., Saratoga Springs, N. Y.
James H. Jackson, M. D., Dansville, N. Y.
K. J. Jackson, M. D., Dansville, N. Y.
P. W. Neefus, M. D., Rochester, N. Y.
James K. King, M. D., Watkins, N. Y.
Jefferson Scales, M. D., New Brighton, N. Y.
Seth S. Goldthwaite, M. D., Boston, Mass.
L. S. Dixon, M. D. (Harvard), Boston, Mass.
T. M. Strong, M. D., Boston, Mass.
John J. Shaw, M. D., Plymouth, Mass.

Sanford Hanscom, M.D., E. Somerville, Mass.:
"And without pain."

Ernest N. Noyes, M. D., Newburyport, Mass.
David Clark, M. D., Springfield, Mass.
John Sanborn, M. D., Melrose, Mass.
B. F. Moulton, M. D., Lawrence, Mass.
George M. Morse, M. D., Clinton, Mass.
C. C. Cundall, M. D., Fairhaven, Mass.
Daniel Humphrey, M. D., Lawrence, Mass.
Herbert F. Pitcher, M. D., Haverhill, Mass.
Wm. Winslow Eaton, M. D. (Univ. of N. Y.),
Danvers, Mass.
S. K. Merrick, M. D., Boston, Mass.
William H. Carpenter, M. D., Boston, Mass.
Benjamin H. Hartwell, M. D., Ayer, Mass.
Walter Channing, M. D., Brookline, Mass.
John Homer, M. D., Newburyport, Mass.
Wm. H. Milliken, M. D., Boston, Mass.
Camille Coté, M. D., Marlboro', Mass.
Charles N. Page, M. D., Danvers, Mass.
Frank E. Bundy, M. D., Boston, Mass.: —
"And without pain, when practicable."
W. F. Wesselhoeft, M. D., Boston, Mass.
Elisha Chenery, M. D., Boston, Mass.: —
"II. and III. should go together."
Wm. P. Stutson, M. D., Cummington, Mass.
Sarah Hackett Stevenson, M. D., Chicago, Ill.
Rachel H. Carr, M. D., Chicago, Ill.
Almon Brooks, M. D., Chicago, Ill.
George M. Palmer, M. D.
W. B. Lewett, M. D., San Francisco, Cal.
Charles W. Stockman, M. D., Portland, Me.
Rev. Dr. Wilford L. Robbins, Albany, N. Y.
Rev. O. B. Frothingham, Boston, Mass.
Rev. Dr. Horatio Stebbins, San Francisco, Cal.
Rev. Dr. B. L. Agnew, Philadelphia, Pa.
Rev. R. Leroy Lockwood, Broomfield, N. J.
Rev. Henry Blanchard, Portland, Me.
Rev. M. J. Savage, Boston, Mass.
Rev. Dr. William Salter, Burlington, Ohio.
Rev. Charles F. Dole, Boston, Mass.
Rev. Dr. Thomas K. Beecher, Elmira, N. Y.
Rev. Dr. H. W. Thomas, Chicago, Ill.: —
"I favor this statement; but as far as
possible vivisection should be painless."
Rev. Dr. Thomas B. Angell, Harrisburg, Pa.:
"The second statement most nearly represents my ideas, although the reasons
advanced in the third statement and
the limitations therein suggested also
strongly appeal to me. Regarding
vivisection as a method of study in
some public and private schools, I hold
the strongest opinions; namely, that
such use should be utterly, entirely,
and definitely prohibited under the
extremest penalties."

VIVISECTION RESTRICTED BY UTILITY.

Rev. Dr. Howard A. Johnston, Chicago, Ill.

Rev. Dr. Thomas C. Hall, Chicago, Ill.

Rev. William W. Jordan, Clinton, Mass.

Rev. Dr. Wm. R. Richards, Plainfield, N. J.:
" This statement commends itself to my judgment. . . . I am sometimes made uneasy by what seems to me a tendency at present towards excessive control of individual liberty by law."

Rev. Dr. William Bryant, Editor of the "Michigan Presbyterian," Mt. Clemens, Michigan: —
" No. III. most nearly meets my views, but I see some points in I. and II. with which I also agree. I question how far vivisection is valuable."

Rev. Dr. J. De Hart Bruen, Belvidere, N. J.:
" I appreciate the great difficulty in making restrictions ; yet a reasonable law, I believe, could be drawn and enforced which would check abuse, and yet not seriously embarrass the search for useful facts."

Rev. Alfred Noon, Ph.D., Boston, Mass.:
" While on the whole favoring this presentation, I sign with two expressions of modification: 1. The utility idea should be modified by the provision of painlessness. 2. Am not clear that the State is the best repository of power. In many communities the local government could enforce better than the wider constituency."

Rev. Charles H. Walker, Lansingburgh, N. Y.
(Adds final paragraph of statement II. to this statement.)

Rev. Dr. William H. Davis, Detroit, Mich.

Rev. H. L. Mitchell, Ph.B., Mystic, Conn.:
" While I should be glad to see all infliction of unnecessary suffering upon our dumb friends abolished, yet I think that the restriction of the practice of vivisection by law is the most practicable way to diminish the evil at present."

Rev. Dr. Edward B. Goodwin, Chicago, Ill.:
" I believe most heartily in the restrictions indicated under this head, and I would have the law of restrictions most rigidly applied, and its penalties rigidly enforced."

Rev. Dr. Isaac J. Lansing, Boston, Mass.:
" I favor this, with a leaning toward No. I. It is solely a question whether men can be saved from suffering by inflicting the least possible suffering on lower creatures."

Rev. Dr. W. M. Paden, Philadelphia, Pa.

Rev. Dr. T. Romeyn Beck, Oakland, Cal.

Rev. Dr. George M. Steele, LL.D., Ex-Pres. Lawrence University, Auburndale, Mass.

Rev. Dr. Rufus A. White, Chicago, Ill.: —
" This my judgment signs; my sympathies sign the first."

Rev. Dr. John McClellan Holmes: —
" I am disposed to annex to this statement the views expressed in No. II."

Rev. John W. Chadwick, Brooklyn, N. Y.:
" I heartily approve of at least so much restriction as is here indicated."

Rev. Dr. E. Winchester Donald, Boston, Mass:
" Permission to practise vivisection should be granted only by a license issued by a competent board. Vivisection without such a license should be made a punishable offence."

Rev. Dr. John Henry Burrows, Chicago, Ill.
(Dr. Burrows would make the seventh line stronger by inserting " very narrow " before the word " limitations.")

Rev. Dr. Thaddeus A. Snively, Chicago, Ill.

Rev. J. E. Roberts, Kansas City, Mo.

Rev. Wm. H. Clark, Bay City, Mich.: —
" Without pain whenever possible."

Rev. S. B. Alderson, D. D., Topeka, Kan.
(Erases last two paragraphs.)

Rev. Marcus N. Preston, Bath, N. Y.

Rev. L. H. Hallock, D. D., Tacoma, Wash.

Rev. S. H. Cobb, Richfield Springs, N. Y..

Rev. V. L. Lockwood, D. D., Bloomfield, N. J.

Rev. Joseph Osgood, Cohasset, Mass.

Rev. Louis C. Washburn, Rochester, N. Y.

Rev. Samuel C. Palmer, St. Louis, Mo.

Rev. A. A. Kiehle, D. D., Milwaukee, Wis.

Rev. J. M. Seymour, Norwalk, Ohio.

Rev. L. R. Dalrymple, Reading, Pa.

Rev. Arthur C. Powell, Baltimore, Md.

Rev. D. W. Coxe, Archdeacon of Scranton, Pa.

Rev. S. W. Derby, Rockville, Conn.

Rev. C. P. Anderson, Oak Park, Ill.

Rev. Frederick L. Hosmer, St. Louis, Mo.

Rev. William B. Clark, Seneca Falls, N. Y. :
" Utility to be unquestionable, and vivisection to be without pain so far as possible."

Rev. I. W. Hathaway, D. D., Jersey City, N. J.

Rev. Fred T. Rouse, Plantsville, Ct.

Rev. John Rouse, M. A. (Oxon.), Chicago, Ill.

Rev. Edward Abbott, D. D., Cambridge, Mass.

Rev. Henry H. Stebbins, Rochester, N. Y.

Rev. Cornelius W. Morrow, Norwich, Conn.

Rev. James W. Cooper, New Britain, Conn.

. 4

Rev. John Acworth, New York City.
Rev. Walter M. Barrows, D. D., Rockford, Ill.
Rev. Charles Townsend, Cleveland, Ohio.
Rev. Henry L. Jones, Wilkesbarre, Pa.
Rev. Thomas L Cole, Portland, Oregon.
Rev. Arthur L. Williams, Chicago, Ill.
Rev. W. R. Huntington, D. D., Rector of Grace Church, New York City.
Rev. C. D. W. Bridgman, New York City.
Rev. W. J. Petrie, Chicago, Ill.
Rev. Edwin P. Thomson, Springfield, Ohio.
Rev. Judson Titsworth, Milwaukee, Wis.
Rev. Frank Russell, Bridgeport, Conn.
Rev. C. S. Nickerson, Racine, Wis.
Rev. Albert W. Ryan, Duluth, Mich.
Rev. Samuel B. Stewart, Lynn, Mass.
(Erases last line but one, and inserts the stronger phraseology of the last four lines of II.)

Rev. Clinton Douglas, Des Moines, Iowa.
Rev. S. C. Beach, Bangor, Me.
Rev. Austin B. Bassett, B. D., Ware, Mass.
Rev. James McLeod, D. D., Scranton, Pa.
Rev. Percy Browne, Boston, Mass.
Rev. Arthur W. Little, D. D., Evanston, Ill.
Rev. J. L. Parks, S. T. D., Philadelphia, Pa.
Rev. C. H. Hamlin, Easthampton, Mass.
Rev. Theo. B. Foster, Pawtucket, R. I.
Rev. P. N. Meade, Oswego, N. Y.
Rev. Calvin M. Clark, Haverhill, Mass.
Rev. Edward C. Ewing, Danvers, Mass.
Rev. Leon P Marshall, Franklin, Ind.
Rev. James M. Patterson, Detroit, Mich.
Rev. William W. Knox, D. D., New Brunswick, N. J.: —
"With the least possible pain."

Rev. Irving W. Metcalf, Cleveland, Ohio.
Rev. Frederick M. Kerkees, Meadville, Pa.
Rev. Henry H. Sleeper, Ph.D., Elizabeth, N.J.
Rev. D. A. Newton, Winchester, Mass.
Rev. George F. Smythe, Mount Vernon, Ohio.
Rev. G. A. Alcott. M. A., Danielsonville, Ct.
Rev. Edward McArthur Noyes, Newton Centre, Mass.
Rev. James H Van Buren, Lynn, Mass.
Rev. Emery H. Porter, Newport, R. I.
Rev. John H. Egar, D. D., Rome, N. Y.
Rev. John H. Griffith, Albany, N. Y.
Rev. H. S. Rablee, Charlotte, Mich.
Rev. Russell T. Hall, Greenwich, Conn.
Rev. M. D. Edwards, D. D., St. Paul, Minn.
Rev. J. Spencer Voorhees, M. A., West Winsted, Conn.
Rev. S. M. Crothers, Cambridge, Mass.
Rev. Raymond H. Stearns, Bainbridge, N. Y.
Rev. John P. Peters, New York City.
Rev. Thomas H. Sill, New York City.

Rev. Joseph Gambee, Plattsburgh, N. Y.
Rev. D. L. Sanford, Bellows Falls, Vt.
Rev. B. W. Lockhart, Manchester, N. H.
Rev. W. L. Phillips, D. D., New Haven, Conn.
Rev. George A. Paull, Bloomfield, N. J.
(Adds last clause of II.)

Rev. Frederick Gibson, D. D., Baltimore, Md.
Rev. William Lloyd Himes, Concord, N. H.
Rev. Charles Herr, S. T. D., Jersey City, N. J.
Rev. Albert M. Hilliker, Washington, D. C.
Rev. G. S. Mott, D. D., Flemington, N. Y.
Rev. William Ballou, Fargo, N. D.
Rev. J. R. Collier, Louisville, Ky.
Rev. Andrew C. Browne, Peoria, Ill.
Rev. George E. Martin, St. Louis, Mo.
Rev. Caroline J. Bartlett, Kalamazoo, Mich.
Rev. William Dallmann (Editor of the "Lutheran Witness"), Baltimore, Md.
Rev. David Magie, Paterson, N. J.
Rev. Charles T. Haley, D. D , Newark, N. J.
Rev. A. Gosman, Lawrenceville : —
"But practised only by scientific men and for clearly scientific ends."

Rev. Wayland Spaulding, Poughkeepsie, N.Y.
Rev. George O. Little, Washington, D. C.
Rev. T. J. Lamont, M. A., Olympia, Wash.
Rev. W. F. Paddock, D. D., Philadelphia, Pa.
Rev. Dr. J. A. Pollock, Lebanon, Ind.
Rev. W. E. Allen, Sherburne, N. Y.
Rev. Wm. R. Mulford, New Haven, Conn.
Rev. William R. Campbell, Roxbury, Mass.
Rev. Christopher G. Hazard, Catskill, N. Y.
Rev. Amos Skeele, Rochester, N. Y.
Rev. E. L. House, S. T. B., Attleboro, Mass.
Rev. A. McCullagh, D. D., Worcester, Mass.
Rev. Willard B. Thorp, Binghamton, N. Y.
Rev. Robert E. Ely. Cambridgeport, Mass.
Rabbi Abraham R. Levy, B. Ph., Chicago, Ill.
Rev. James Roberts. D. D., Colwyn, Pa.
(Accepts II. as best expression of his views while not endorsing its entire phraseology.)

Rev. Dr. A. S. Fiske, Ithaca, N. Y.: —
"I believe this practice should be regulated by intelligent laws, restricted to competent and authorized persons and fit places, and conducted solely in the interests of relief to human pains and diseases."

Rev. Fred. E. Dewhurst, Indianapolis, Ind.
Rev. George Hutchinson Smith, D. D., East Orange, N. J.
Rev. Samuel Cedsall, Chicago, Ill.
Rev. Dr. John Acworth, New York City.
Rev. Clifford W. Barnes, B. D., Chicago, Ill.
Rev. Dr. Edmund Q. S. Osgood, B. D., Hyde Park, Mass.

Rev. Dr. Myron Adams, Rochester, N. Y.

Rev. Nathaniel Seaver, Jr., Leicester, Mass.:
"I would confine vivisection to medical and scientific schools, and prohibit it elsewhere altogether, especially in presence of minors."

Rev. A. S. Garver, D. D., Worcester, Mass.

Rev. T. M. Hodgdon, West Hartford, Conn.:
"Vivisection should be without pain so far as possible."

Rev. Dr. Wm. A. Keese, Lawrence, Mass.:
"I am by no means sure but that the first statement is correct; yet not having special knowledge on the subject, I incline to the more conservative opinions of this statement."

Rev. Dr. C. E. S. Rasay, Carthage, N. Y.:
"I should favor absolute prohibition of vivisection did I not think that there are times when its practice results in good so great as to justify the sacrifice of the animal. Undoubtedly these cases are rare, and restrictions should be made even stronger than indicated above."

Rev. Dr. T. T. Munger, New Haven, Conn.:
"Vivisection should be under strict and careful supervision, but with a leaning to No. II. I would confine it to experts for the sake of discovery, and shut it out as means of demonstrating established facts."

Rev. Wm. C. Stiles, D. D., Jackson, Mich.:
"Vivisection should always be without pain when anæsthetics do not prevent the experiment."

Rev. Joseph J. Woolley, Pawtucket, R. I.

Rev. John G. Davenport, S. T. D., Waterbury, Conn.

Francis Ellingwood Abbott, Ph.D., Cambridge, Mass.:—
"The above statement comes the nearest to expressing my views on the subject, but I object strongly to the word 'utility.' Vivisection is a human action, a part of conduct, and all conduct must be governed by *ethical principle*, not mere calculation of consequences. Vivisection, if really conducive to knowledge (of which I am no judge), may be prompted by motives of the purest moral character, by the purpose of inflicting *some* pain to prevent greater pain, as in surgical or dental operations. Only as a means to mercy and ultimate diminution of suffering ought vivisection to be tolerated."

Rev. S. J. Smith, Ph.D., St. Paul, Minn.

Rev. George T. Linsley, Newtown, Conn.

Rev. H. R. Lockwood, S. T. D., Syracuse, N.Y.

Rev. E. C. Murray, D. D., Clinton, S. C.

Rev. A. C. Kimber, S. T. D., New York City.

Rev. Sam'l Scoville, D. D., Stamford, Conn.:
"With a very strong leaning toward No. II. and looking for the time when No. I. shall be the law."

Rev. Barton O. Aylesworth, LL.D., President of Drake University, Des Moines, Iowa.

Edwin L. Godkin, Editor of "The Nation," New York City.

W. J. Rolfe, Ph.D., Editor and Author, Cambridge, Mass.

Carlos Martyn, LL.D., Editor, Chicago, Ill.

F. B. Sanborn, Journalist, Concord, Mass.

Hjalmar H. Boyesen, Author; Professor at Columbia College, New York.

Nathan H. Dole, Literarian, Boston, Mass.:
"Under only the most stringent restrictions!"

G. W. Turner, Editor, New York City.

Geo. Wm. Winterburn, M. D., Editor, New York City.

A. Ludlow White, Editor, New York City.

W. C. Dunn, Editor, New York City.

John Y. Foster, Editor, New York City.

E. J. Wheeler, A. M., Editor, New York City.

Sam'l T. Pickard, Editor, Portland, Me.

H. L. Hastings, Editor, Boston, Mass.

E. F. Hartshorne, Editor, Boston, Mass.

A. W. Stevens, Editor, Cambridge, Mass.

Louise M. Hodgkins, Editor, Boston, Mass.

Ernest E. Russell, Editor, New York City.:
"No experiments involving pain should be permitted in the class-room. Cases are conceivable, however, where such experiments may be justified, if conducted as humanely as possible, and with due sense of the grave responsibility assumed by the investigator, and of his moral obligation to restrict the suffering of the creature experimented upon within the narrowest bounds consistent with the attainment of the object sought."

Albert G. Boynton, Editor, Detroit, Mich.

Amos R. Wells, Editor, Boston, Mass.

Ernest R. Willard, Editor, Rochester, N. Y.

John Lemley, Editor, Albany, N. Y.

Henry Abbott Steele, Editor, Newark, N. J.

James H. Potts, Editor, Detroit, Mich.

Stephen Quinon, Journalist, Pittsburg, Pa.

T. J. Keenan, Jr., Editor, Pittsburg, Pa.

Frank Kasson, Editor of "Education," Boston, Mass.

R. S. Thompson, Editor, Springfield, Ohio.
C. A. Clardy, M. D., Newstead, Ky.
Gov. Wm. A. Richards, Cheyenne, Wyo.
Gov. J. M. Stone, Jackson, Miss.
Gov. Claude Mathews, Indiana.
Gov. D. Russell Brown, Providence, R. I.
Gov. John Gary Evans, Columbia, S. C.
Hon. J. A. T. Hull, M. C., Des Moines, Iowa.
Hon. J. T. M. Cleary, M. C., Mankato, Minn.
Hon. R. W. Tayler, M. C., New Lisbon, Ohio.
Hon. Alex. M. Hardy, M. C., Washington.
Hon. Matthew Griswold, M. C., Erie, Pa.
Hon. F. C. Layton, M. C., Wapakoneta, Ohio.
Hon. C. N. Fowler, M. C., Elizabeth, N. J.
George C. Holt, Esq., New York City: —
  " The law should define the general classes
  of cases in which vivisection should be
  allowed, and medical officials should
  issue a license in advance for the pro-
  posed experiments, clearly fixing the
  nature of them."

Hon. C. F. Buck, M. C , New Orleans, La.:
  " Regulation should be universal; the civil-
  ized countries should unite action."
F. R. Coudert, Esq., New York City.
Hon. Binger H. Hermann, M. C., Washing-
  ton, D. C.
James R. Howe, M. C., Brooklyn, N. Y.
Hon. Frank M. Eddy, M. C., Glenwood, Minn.
Hon. W. C. Owens, M. C., Georgetown, Ky.
Hon. Charles A. Towne, M. C. (Ph.B.), Du-
  luth, Minn.
Hon. W. M. Denny, M. C., Scranton, Miss.
Hon. H. M. Baker, M. C., Bow Mills, N. H.
Hon E. J. Murphy, M. C., East St. Louis, Ill.
Hon. W. K. Ellis, M. C., Heppner, Ore.
Hon. I. P. Wanger, M. C., Norristown, Pa.
Hon. A. B. Wright. M.C., North Adams, Mass.
Rev. William M. Salter, Philadelphia, Pa.
B. De Witt, M. D., Oswego, N. Y.
George E. Mucuen, M. D., Boston, Mass.
Henry W. Boorn, M. D., Schenevus, N. Y.

2. Signers accepting the main Propositions, but altering Phraseology.

CHARLES ELIOT NORTON, LL.D., Harvard University, Cambridge, Mass.

JOHN H. GLADSTONE, D. Sc., F. R. S., London, England.

ASAPH HALL, Jr., Professor of Astronomy, University of Michigan, Ann Arbor.

C. R. VAN HISE, Professor of Geology, University of Wisconsin, Madison: —

  (Signs last two paragraphs.) " It is not wise to make restrictions based upon known utility."

Rev. JOHN J. McCOOK, Trinity College, Hartford, Conn. : —

  " There are not probably many men who would conduct experiments involving pain or loss of life without using an anæsthetic, unless there were some real and convincing necessity ; and yet I have witnessed some. I fear it is a fact that vivisection inevitably tends to make the conscience of the operator a trifle too easy in regard to the whole matter. Restraint ought to be carefully limited, however."

Prof. A. D. F. HAMLIN, Columbia College, New York City.

Prof. D. BODLIGHT, North Western University, Evanston, Ill.

ROSSITER W. RAYMOND, New York City.

SIR JOSEPH FAYRER, M. D., K. C. S. I., Surgeon General, London :

  (Sir J. Fayrer strikes out several words and clauses, particularly the references to Magendie and Mantagazza. He leaves, however, the follow-lowing sentences intact :)

  " We regard as cruel and wrong the infliction of torment upon animals in the search for physiological facts which have no conceivable relation to the treatment of human diseases. . . . We consider as wholly unjustifiable the practice of subjecting animals to torture in the laboratory or class-room merely for the purpose of demonstrating well-known and accepted facts. . . . Such experiments as these are degrading in tendency."

(Reference is often made to Sir J. Fayrer's numerous experiments in India on snake-poison; and his indorsement of the foregoing paragraphs is the more striking as coming from one of the leading experimenters.)

MALCOLM MCLEAN, M. D., Surgeon-in-charge, St. Andrew's Infirmary, New York City : —

" . . . Vivisection should be limited to actual necessity of such research, and not be so generally practised for demonstration only."

Hon. SIMEON E. BALDWIN, New Haven, Conn. : —

" Restricted by utility and humanity."

Prof. E. I. JAMES, University of Pennsylvania, Philadelphia, Pa.
A. T. BRISTOW, M. D., Surgeon at the Long Island College Hospital, Brooklyn, N. Y. : —

" It is unfair to Science to ask her in all cases to state what she expects to prove by a given line of experimentation. Often, we can only 'try and see.' Nor can one state that any particular physiological fact will *never* bear some relation to the treatment of disease. I am entirely in accord with the third paragraph."

Jerome Walker, M. D., Brooklyn, N. Y.
Herbert F. Williams, M. D., Brooklyn, N. Y.
Robert T. Edes, M. D. (formerly Professor at the Harvard Medical School), Saratoga, New York.
Lewis O. Goetchius, M. D., Saratoga, N. Y.
Julius J. Kempe, M. D., Rochester, N. Y.
W. O. Stillman, M. D., Albany, N. Y.
C. E. Stebbins, M. D., Morris, N. Y.
E. G. Inlay, M. D., Saratoga Springs, N. Y.
G. W. Sargent, M. D., Seneca Castle, N. Y.
Wm. C. Ney, M. D., Elmira, N. Y.
Wm. More Decker, M. D., Kingston, N. Y.
Chas. H. Perry, M. D., Oneida, N. Y.
Jos. Hasbrouck, M. D., Dobbs Ferry, N. Y.
(Apparently signs only last paragraph.)
J. A. Irwin, M. A. (*Cantab.*), M. D., New York City.
John Cabot, M. D., New York City.
Alex. W. Stein, M. D., New York City.
George G. Needham, M. D., New York City.
Wm. A. Valentine, M. D., New York City.
H. J. Garrigues, M. D., New York City.
John W. Elliot, M. D., Boston, Mass.
Salome Merritt, M. D., Boston, Mass.
E. W. Cushing, M. D., Boston, Mass.
Henry C. Angell, M. D., Boston, Mass.
R. C. Macdonald, M. D., Boston, Mass.
W. Thornton Parker, M. D., Groveland, Mass.

Wm. O. Faxon, M. D., Stoughton, Mass.
D. W. Vanderbergh, M. D., Fall River, Mass.
E. B. Cutler, M. D., Waltham, Mass.
J. N. Danforth, M. D., Chicago, Ill.
G. H. Parkhurst, M. D., Brooklyn, N. Y. :

" Utility alone cannot give Science her authority, and that utility which would demonstrate that which we have already repeatedly demonstrated should be excluded. Neither should the infliction of pain be permitted. Vivisection should be permitted to competent and trustworthy persons, and restricted to licensed places which shall be open at all times to inspection by agents of humane societies, by members of the medical profession, or by officers duly appointed or empowered by law."

Susan E. Crocker, M. D., Boston, Mass.
Joseph P. Paine, M. D., Roxbury, Mass.
(Erases "to determine action of new remedies.")

C. E. Banks, M. D., Vineyard Haven, Mass.:
" I sign this without adopting *all* the language or sentiment of it. . . . I agree to the desirability of prohibiting the demonstration of well-known physiological facts before classes."

Francis E. Corey, M. D., Westboro, Mass.
(Erases lines 13–14, as to degrees of pain.)
Sam. B. Woodward, M. D., Worcester, Mass.
J. H. Hobart Burge, M. D., Brooklyn, N. Y.:
"The above statement is a fair presentation of my personal views." (Dr. Burge erases the word "common" in nineteenth line.)
George C. Webber, M. D., Millbury, Mass.:
"I think it quite impossible to say that *any* physiological fact can have ' no relation to the treatment of human diseases.' "
J. T. G. Nichols, M. D., Cambridge, Mass.
James T. Walker, M. D., Falmouth, Mass.
(Erases 4th line to "necessity.")
Mary A. Mixer, M. D., Chicago, Ill.
(Erases lines 13–14, as to degrees of pain.)
Thomas W. Busche, M. D., New York City.
Prof. Francis Valk, M. D., N. Y. Postgraduate College. N. Y.
J. N. Wright, M. D., Grand Gorge, N. Y.
(He says of paragraph III. referring to painful demonstrations, "Especially good. Too much of this is done.")
William F. Dudley, M. D., Brooklyn, N. Y.
(Dr. Dudley would strike out paragraph referring to instantaneous death and the clause which follows.)
Lt.-Col. Alfred A. Woodhull, M. D., LL.D., Deputy Surgeon-General, U. S. A.
(Col. Woodhull erases references to Magendie and Mantagazza and a large number of single words which appear to him superfluous. He indorses, however, the statements condemning repetition of painful experiments for demonstration and favoring legal restrictions.)
Prof. Joseph Ransohoff, M. D., F. R. C. S., Professor of Surgery, Medical College of Ohio, Cincinnati, Ohio.
(Erases first sentence of third paragraph and adds "except under anæsthesia.")
Prof. C. M. Cary, Professor of Physiology Auburn, Ala.

Rev. Hermann Lebenthal, Weathersfield, Ct.:
"With a broader definition of utility."
William Morton Payne, Associate Editor of "The Dial," Chicago, Ill.: —
"'Vertebrates' should be substituted for 'animals' throughout. With invertebrates our sympathies would be wasted."
Pres. Jos. W. Mauck, State University of South Dakota, Vermillion, S. D.
Pres. Samuel W. Boardman, Maryville College, Maryville, Tenn.
Pres. George Sverdrup, Augsberg Seminary, Minneapolis, Minn.
Rev. Dr. W. H. Ward, LL.D., Superintending Editor of "The Independent," N. Y. City.
H. K. Carroll, LL.D., Religious Editor of "The Independent," New York City.
H. L. Wayland, Editor, Philadelphia, Pa.
Samuel Sexton, M. D., New York City : —
"I do not care to indorse some parts of the statement, but you may classify me in a general way as opposed to the abuses of vivisection as they now exist."
Rev. S. B. Pond, Norwalk, Conn.
Rev. S. W. Meek, Peoria, Ill.
Rev. S. P. Wilder, Janesville, Wis.
Rev. S. G. Wood, Easthampton, Mass.
Rev. W. G. Andrews, D. D., Guilford, Conn.
Rev. Charles E. St. John, Pittsburg, Pa.:
"Utility should include laboratory and class-work demonstrations for mature students, but not before children."
Rev. Dr. Henry Van Dyke, New York City.
(Strikes out from the middle of the 6th line to the end of the 2d paragraph.)
Rev. Dr. W. P. Swartz, Wilmington, Del.
Hon. Edgar Wilson, M. C., Boise City, Idaho.
(Erases line referring to "instantaneous death." "In other respects I indorse above statement.")
Hon. Wm. L. Terry, M. C., Little Rock, Ark.
(Erases references to Mantagazza; lines 23–25 and part of lines 27–28. "Should be . . . practise.")

3. Those who condemn Torture, for Teaching, but do not approve of Legislation, nor of any Impediments to Original Research. Other slight Changes in Phraseology are not infrequent.

Prof. James Law, F. R. C. V. S., Professor of Veterinary Science, Cornell Univ., Ithaca, N. Y.: —
"As regards the unnecessary repetition of painful physiological experiment before a class, or even in a laboratory, to illustrate a truth which is already proved, it can only deserve the condemnation of any man of humanity. . . . The infliction of pain for the mere purpose of demonstration is, in my opinion, wholly unwarranted."

Prof. L. H. Bailey, College of Agriculture, Cornell University, N. Y.: —

"I subscribe to III. so far as the fundamental position is concerned, but do not believe in legal measures to control it. I should prefer to control it by an agreement among the leading educators of the country."

Dr. George Trumbull Ladd, Professor of Psychology, Yale College, Conn.: —

"This represents my views most nearly, with the exception of the last clause. I have increasingly little respect for the law-makers of this country on this or any similar subject. I should far rather trust the men who practise vivisection, unfeeling and brutal as many of them undoubtedly are."

S. Thompson, Editor of the "Chicago Evening Journal," Chicago, Ill.: —

"In every case the end should justify the means, and the experimenter should be held to accountability whenever he oversteps what is necessary to his search after knowledge. Nothing should be forgiven to exhibition or curiosity."

C. R. Williams, Journalist, Indianapolis, Ind.
Charles H. Levermore, Ph.D., Brooklyn, N. Y. (Erases last seven lines.)

Rev. Willis J. Beecher, Professor in Auburn Theological Seminary, N. Y.

Prof. James M. Baldwin, Ph.D., Professor of Psychology, Princeton University, N. J.:

". . . The word 'utility' is not to be defined by popular opinion." (Professor Baldwin considers experimentation justifiable in a variety of other circum-

stances than those named in the statement; "for example, one of the subjects requiring extended experiments just now is the physiological accompaniments and effects of PAIN." But this is precisely the field in which Mantagazza has been working and performing the most atrocious experiments without any results of the slightest value.)

Rev. R. W. Brokaw, M. A., Springfield, Mass.
A. T. Cabot, M. D., Boston, Mass.
Morton H. Prince, M. D., Boston, Mass.
J. G. Hubbard, M. D., Holyoke, Mass.
Isaac H. Stearns, M. D., Lynn, Mass.
Angelo O. Squier, M. D., Springfield, Mass.
Edward P. Fowler, M. D., New York City.
Robert J. Wilding, M. D., Malone, N. Y.
Prof. F. B. Gummere, Haverford College, Pa.:

"Surely not State control through our deplorable State legislatures! The question involves intelligence and conscience."

E. Fletcher Ingals, M. D., Chicago, Ill. (Erases last paragraph.)

Prof. James Tyson, M. D., Univ. of Pennsylvania, Philadelphia, Pa.

Homer E. Smith, M. D., Norwich, N. Y.:

"With these provisos: (1) Prohibited in different schools; (2) Limited in medical schools to painless experiments: (3) No restriction whatever in laboratories of private investigators."

Rev. Dr. Cyrus S. Bates, Cleveland, Ohio:

"I would rather trust the final solution of the vivisection question to education and moral elevation than to legislation."

4. Signers who apparently approve of certain Restrictions of Vivisection, but who do not condemn even painful Experiments, if approved by a Professor.

Sir Dyce Duckworth, M. D., LL. D., Lecturer at St. Bartholomew's Hosp., London, Eng.

Prof. John M. Tyler, Prof. of Biology, Amherst College, Mass.: —

"Anæsthetics should be employed whenever possible; I do not think that they can always be employed. I heartily agree with the closing paragraph, that vivisection should be controlled by the law of the State."

George Seymour, M. D., Utica, N. Y.
Rev. James Eells, D. D., Englewood, N. Y.

Albion W. Tourgee, Ph.D, LL.D., Author, Mayville, N. Y.: —

"I do not think it incumbent on Science to prove that vivisection is necessary. It need not even be necessary at all; if it be helpful to the student or operator, then it should be allowed under proper restrictions. . . . I should vigorously oppose any other restriction than that of a well-guarded license." (Erases all but last paragraph of the statement.)

Rev. G. H. Beard, Ph.D., So. Norwalk, Conn.

H. H. Baxter, M. D., Cleveland, Ohio.
(Considers painful experiments justifiable for demonstration, if " not extended to gratify an idle or morbid, or even a scientific, curiosity.")

Thomas M. L. Chrystie, A. M., M. D., New York City.

Prof. F. A. Blackburn, Univ. of Chicago:
" I should be inclined to favor a restriction on persons rather than on methods, — a system of licenses that would authorize a medical professor to practise vivisection at his own judgment as an aid in teaching, but would prevent his students from experimenting *except under his direction.*"

Edward P. Nichols, A. M., Boston, Mass.

Prof. Charles Foster Smith, Professor of Greek, University of Michigan: —
" I have not much confidence in the wisdom of a State or even a National Legislature in dealing with such questions." (As to restriction or non-restriction, he would be guided by the decision of scientific men.)

5. Those whose Alterations nullify the whole Statement.

Rev. Arthur Chase, B. D., Boston, Mass. :
" *To trust the professor wholly,* unless there seemed reason for distrust, expresses to my mind the best policy: vivisection governed by the moral sense."

Prof. C. M. Grumbling, Professor of Biology, etc., Iowa Wesleyan University, Mount Pleasant, Iowa: —
" I lean to the side of freedom on the part of specialists. He will nearly always administer an anæsthetic when operating on the higher animals, unless the case in hand forbids."

A. H. Carvill, M. D., Somerville, Mass.

Alice F. Mills, M. D., Binghamton, N. Y. :
" Let anæsthetics be used whenever they will not defeat the ends of science."

James Gerrie, M. D., Brooklyn, N. Y.:
"Suffering may be necessary to make fact clear to the student's mind."

Rev. Dr. W. Durant, Saratoga Springs, N. Y.
(Dr. Durant erases references to Magendie and Mantagazza, and would permit vivisection where there is *probability* of benefit and the suffering not greater than "necessary to the end in view.")

A. Walter Suiter, M. D., Herkimer, N. Y.

James L. Turner. M. D., New York City:
" The benefit of ocular demonstrations is too great to be ignored for the manifestation of mere sentiment. Utility and necessity should be the governing factors in matters of this kind."

John C. Schapps, M. D., Brooklyn, N. Y.

Prof. Edward D. Cope, Professor of Zoölogy, etc., Univ. of Pennsylvania.
(Prof. Cope considers " wholly *justifiable* the common practice in the United States of subjecting animals to torture for the purpose of demonstrating well-known facts," etc. He would control vivisection by the faculties of the various institutions of learning, wherein it is practised, such as " universities, academies of science, and medical schools." They " should grant licenses to practise vivisection to whomsoever they should deem suitable persons.")

Prof. Amos G. Warner, Economics and Social Science, Stanford Univ., Cal.
(Erases all but the first sentence of the last paragraph.)

Pres. George S. Burroughs, D. D., LL.D., Wabash College, Crawfordsville, Ind.
(Dr. Burroughs does not believe in State control, nor in legal abolition of painful experiments.)

S. G. Shank, M. D., Albany, N. Y.

James L. Perry, M. D., New York City.

H. A. C. Anderson, M. D., New York City.

A. A. Hubbell, M. D., Buffalo, N. Y.

Z. Edwards Lewis, M. D., New Rochelle, N. Y.

Edward T. Williams, M. D., Boston, Mass.

A. Lawrence Mason, M. D., Boston, Mass.

O. T. Wilsey, M. D., Amityville, N. Y.: —
" I believe in unrestricted vivisection by scientific investigators, duly licensed."

George H. Weaver, M. D., Prof. of Pathology in Northwestern Univ., Chicago, Ill.

IV (A). **VIVISECTION WITHOUT RESTRICTIONS.**

VIVISECTION, or experimentation upon living creatures, must be looked at simply as a method of studying the phenomena of Life. With it morality has nothing to do. It should be subject neither to criticism, supervision, nor restrictions of any kind. It may be used to any extent desired by any experimenter (no matter what degree of extreme or prolonged pain it may involve) for demonstration before students of the statements contained in their textbooks, as an aid to memory; for confirmation of theories; for original research; or for any conceivable purpose of investigation into vital phenomena. We consider that sentiment has no place in the physiological laboratory; that animals have there no "rights" which Man is called upon to notice or respect; that Science cannot be "cruel" when her sole purpose is to investigate or demonstrate; that it is as great an impertinence for Religion or Morality to assume to sit in judgment upon a scientific method, or to dictate to physiologists limitations beyond which extreme pain "*ought not* to be inflicted," as for Theology to tell the astronomer where in the skies he should not direct his telescope, or the geologist what rocks he must not break.

And finally, while we claim many discoveries of value in the treatment of human ailments to have been due to experiments upon animals, yet even these we regard as of secondary importance to *the freedom of unlimited research, and the independence of Science from all restrictions or restraints*, — agreeing largely with the statement of Dr. Hermann, Professor of Physiology at Zurich, that "no true investigator in his researches thinks of their practical utilization, *that the advancement of Knowledge*, and not practical utility to Medicine, is the true and straightforward object of all Vivisection."

President J. G. SCHURMAN, Cornell University, Ithaca, N. Y.

President DAVID S. JORDAN, M. D., Ph.D., LL.D., Stanford University, California : —

"I believe that any attempt to restrict vivisection in the hands of competent men, properly fitted to act as investigators, would be thoroughly mischievous in its results. Nor do I believe that any considerable amount of wanton pain has ever been inflicted by men whose original work as scientific investigators entitles them to the name. On the other hand, I greatly deprecate the mischief done by persons who are not investigators, but who imitate the actions of these for highly different purposes. The whole matter is one for public opinion to regulate rather than the State. I do not find my views expressed fully in any one of the four statements in the circular; but if I should sign any one of the four it would be this."

Prof. JOSEPH LE CONTE, Professor of Geology and Natural History, University of California, Berkeley : —

"I have carefully examined your propositions in regard to vivisection, and cannot fully endorse any of them; but in the present condition of things in this country I believe I would rather sign this than any of the others. The question in my mind is between the third and fourth. My objection to the fourth is, that the mode of statement is too extreme. . . . I believe licenses should be given (but somewhat freely) to competent and responsible persons, and then no restrictions on mode of experimentation. I believe also it should be used for investigation, not for class-room demonstration.

My objection to the third statement is the use of the word 'utility.' I firmly believe that vivisection should be used — and will always be mostly used — for purely scientific purposes, without any reference to immediate and visible utility. Science will not advance unless truth is sought mainly for its own sake."

(The italics are not in the original, but are here used to point out a sentence which, slightly changed, is incorporated in the next statement.)

Prof. A. E. DOLBEAR, Professor of Physics, Tufts College, Mass.
Prof. ALPHONSE N. VAN DAELL, Mass. Inst. of Technology, Boston.

(Professor van Daell erases all of the second sentence, and the word "simply" in the first sentence. Vivisection should be without restrictions, "at least by persons unfamiliar with scientific methods, and incapable of appreciating the value of experiment. The necessity or use of experiments cannot be under anybody's but the professor's control.")

Prof. THEODORE S. WOOLSEY, Yale Law School, New Haven, Ct. :

"I would advocate vivisection unrestricted except as to the class of experimenters engaging in it. These should be of education and attainments, such as to warrant work of real scientific value. Perhaps a State Commission of competent scientific men could issue licenses."

Dr. BASHFORD DEAN, Instructor in Biology, Columbia College, New York.

Prof. H. S. MUNROE, E. M., Ph.D., Columbia College, New York.

Prof. J. U. NEF, Ph.D. (Munich), University of Chicago, Chicago, Ill.

Prof. I. P. ROBERTS, Cornell University, Ithaca, N. Y.

Prof. W. F. McNUTT, M. D., M. R. C. S., Edinboro', University of California, San Francisco.

Prof. JOHN HENRY GRAY, Ph.D. (Halle), Northwestern University, Evanston, Ill.

(Professor Gray does not wholly agree with this statement, but fails to point out the special clauses to which objection is made. He adds : "I believe there should be restrictions, but not such as described.")

Prof. ARTHUR FAIRBANKS, Ph.D., Yale Divinity School, New Haven, Conn. : —

(Erases second sentence.) "In colleges and universities vivisection should be without restrictions other than those imposed by the instructors."

Prof. A. D. Hurt, LL.D., Tulane University, New Orleans, La.

Prof. Jere. W. Jencks, Ph.D., Cornell University, Ithaca, N. Y.:

"While the statement does not accurately express my views, it does so more nearly than any of the others."

Prof. C. R. Barnes, Ph.D., University of Wisconsin, Madison.

(Professor Barnes erases the word "criticism" in the third sentence; inserts "reputable" before "experimenter;" erases the clause following, referring to degree of pain, and from beginning of fifth sentence to end of first paragraph.)

Prof. S. P. Winans, Ph.D., Princeton University, N. J.: —

"Vivisection should be without arbitrary, legal restrictions."

Prof. Geo. L. Burr, Professor of History, Cornell University, Ithaca, N. Y.: —

"I can by no means assent to either of the four statements; but as my conclusion would be this one, I venture to sign it."

Prof. George C. Comstock, Director of Washburn Observatory, University of Wisconsin, Madison: —

"Vivisection should be without legal restrictions, until the evils of the practice shall become more pronounced than they are at the present time."

Prof. F. Angell, Professor of Psychology, Stanford University, California.

Prof. C. G. Gilbert, Ph.D., Professor of Zoölogy, Stanford University, California.

Prof. J. M. Stillman, Ph.D., Professor of Chemistry, Stanford University, Calfornia.

John Fiske, M. A., LL.D., Author, Cambridge, Mass: —

"I would prohibit vivisection in class-rooms. I agree with the third paragraph of your statement, 'Vivisection restricted by utility.'"

Dr. Paul Carus, Editor of the "Open Court," Chicago, Ill.

Prof. Henry Gradle, M. D., Northwestern University; formerly in the chair of Physiology.

(Does not entirely endorse phraseology.)

Seneca D. Powell, M. D., Professor of Surgery, N. Y. Post-Graduate Medical School, New York City.

Prof. E. Wyllys Andrews, M. D., Professor of Surgery, Chicago, Illinois: —

"Appreciating the high motives of those who would interfere, I nevertheless think each preceptor has a right to be his own guide in such matters, and should be entirely unmolested."

Prof. DANIEL R. BROWER, M. D., Professor of Materia Medica and Therapeutics, Rush Medical College.

Prof. FLEMING CARROW, M. D., University of Michigan.

> (Prof. Carrow erases the words included in parentheses, and the eighth line in the statement.)

Prof. HERMAN KNAPP, M. D., New York College of Physicians and Surgeons, New York City.

Prof. BARTON COOKE HIRST, M. D., Professor of Obstetrics, University of Pennsylvania.

GEORGE F. MORRIS, M. D., late Professor of Physiology, New York City.

Prof. JAMES B. HERRICK, M. D., Chicago, Ill. : —

> "Without restrictions save those imposed by the teacher or professor."

CONRAD WESSELHOEFT, M. D., Professor of Pathology and Therapeutics, Boston University School of Medicine, Boston, Mass. : —

> "Without restrictions, except those of the moral sentiment of sympathy."

A. R. WRIGHT, M. D., Buffalo, N. Y. : —

> "Limit operations to responsible Professors in Laboratory, or oblige every one to obtain a permit from a central authority."

E. BENJAMIN RAMSDELL, M. D., New York City.

SAMUEL B. WARD, M. D., Professor of Theory and Practice of Medicine, Albany Medical College, etc., Albany, N. Y.

MALCOLM LEAL, M. D., New York City : —

> "For purpose of investigation but not for class demonstration, except in special cases where pain is slight and not prolonged."

CHARLES JEWETT, A. M., M. D., Consulting Physician to Kings Co. Hospital, Professor of Obstetrics, Long Island College Hospital, etc., Brooklyn, N. Y. : —

> "Vivisection, or experimentation upon living creatures, may be used to any extent necessary to the advancement of science for original research."
> (Professor Jewett erases all the remainder of the statement, and adds :)
> "*The propositions are all so formulated that none but a fanatic could subscribe to any one of them.*"

SIGISMUND WATERMAN, M. D. (Yale), New York City : —

> "Your mode of procedure deserves hearty success."

With sixty-four other medical men, two educators, and three statesmen.

IV (n). VIVISECTION WITHOUT RESTRICTIONS.

[As the uncompromising phraseology of the first statement seemed objectionable to many, a new statement was formulated toward the close of the inquiry, and sent, with preceding forms, principally to members of the medical profession.]

VIVISECTION, or experimentation upon living animals, must be looked at simply as a method of studying the phenomena of Life; and as such it should be subject neither to criticism, supervision, nor restraints of any kind.

It may be used by any scientific experimenter to any extent he may desire, for demonstration before students of the statements contained in their text-books, as an aid to memory, or for any conceivable purpose of investigation into vital phenomena. And, while many discoveries of value in the treatment of human ailments have undoubtedly been due to experiments on animals, yet even these we regard as of secondary importance to the absolute freedom of research, and the independence of Science from all restrictions and restraints. Pure Science, which exists for its own sake, stands on a far higher plane than Science which exists merely for utility to mankind. We firmly believe that Vivisection should be used, and will always be mostly used, *to add to pure scientific knowledge as such*, *without reference to any usefulness foreseen*. Truth should be sought for its own sake. Dr. Hermann, Professor of Physiology at Zurich, has well said that "no true investigator in his researches thinks of their practical utilization; that the advancement of Knowledge, and not practical utility to Medicine, is the true and straightforward object of all Vivisection."

We would not avoid the question of Pain. It is often the necessity of Vivisection. But Nature will not yield all her secrets without a wrench. For instance, if only by causing acute suffering a teacher can illustrate the functions of the nervous system, should he, merely for that reason, stay his hand ? Ought we to insist that an enthusiastic experimenter should forego any phase of research whatever, simply because of the torture that research may perhaps require ? Such questions afford but one reply. Science does not place reverence for Pity higher than its reverence for any new fact whatever.

In our judgment this question of Pain should be left absolutely to the decision of the experimenter himself. He alone can determine what degree of pain he needs to inflict for the success of his experiment. No laws should constrain him, no critics judge him.

Roswell Park, M. D., Professor of Surgery, University of Buffalo, N. Y.

Francis H. Stuart, M. D., Obstetrician to Brooklyn Hospital, Lecturer on Surgery, Long Island College Hospital, N. Y. : —

" I am in favor of vivisection without other restriction or restraint than the conscience of the experimenter himself."

Nicholas Senn, M. D., Surgeon, Chicago, Ill.

Henry M. Lyman, M. D., Chicago, Ill.

J. H. Raymond, M. D., Professor of Physiology, Long Island College Hospital, Brooklyn, N. Y.

J. Pohlman, Professor of Physiology, University of Buffalo, N. Y. :

" No incompetent or sentimental critics should judge him."

Frederick J. Nott, A. M., M. D., New York City : —

" Whatever restrictions are desirable may be best formulated by the schools and institutions in and for which vivisection is principally done."

William J. Cronyn, M.D., Milwaukee, Wis. : —

" . . . Dr. Bigelow's declaration that ' vivisection deadens the humanity of the students ' may possibly be true in a very limited number of cases, but only in students whose humanity was a minimum to begin with. There are Jesse Pomeroys that are not in prison ; a very few may be in the medical profession ; yet even in these cases it is better that their minimum of humanity be deadened, than that the world should through unnecessary ignorance be deprived of many who have a great deal of it."

Mary Putnam Jacobi, M. D., New York City.

Abby Leach, Professor of Greek. Vassar College, Poughkeepsie, N.Y.

I. Adler, M. D., Professor of Clinical Pathology, N. Y. Polyclinic, New York City.

S. Oscar Myers, M. D., Mt. Vernon, N. Y. : —

" Although there may be an occasional inhuman brute, I think the subject may be safely left without legal restrictions."

Herbert M. Hill, Ph.D., Professor of Chemistry, University of Buffalo, N. Y.

Israel C. Russell, Professor of Geology, University of Michigan, Ann Arbor.

Daniel Laferte, M. D., Professor of Anatomy, Detroit College of Medicine, Mich.

Henry Sewall, M. D., Professor of Physiology, Denver Medical College, Col.

Edward C. Pickering, Astronomer, Cambridge Observatory : —

" Approved (but not if painful) to illustrate known facts."

With one hundred and fifty-three physicians, and three educators.

EXTRACTS FROM LETTERS.

Dr. T. LAUDER BRUNTON, M. D., LL.D., F. R. S., England : —

" I feel very strongly that, while restriction by law is unadvisable and likely to prove harmful, every operator must be, and ought to be, influenced by public opinion and by his own conscience. I hold that vivisection operations belong to the same category as the use of the whip by coachmen ; and that while any instances of abuse, either of experiments or of the whip by the coachman or car-driver, ought to be taken cognizance of, and if necessary punished, it is as objectionable to limit an experimenter in what he is going to do as it would be to pass a law that no driver of a horse or other animal was to carry a whip or other instrument which could be used for the infliction of pain upon the animal he is driving."

(The advocates of restriction of vivisection would probably agree with Dr. Brunton that vivisection operations belong to the same category as the use of the whip by coachmen. They would add, however, that just as the law recognizes what is cruelty on the part of the driver, and what may be permissible to him, it should do the same in the case of the physiologist.)

Sir JOHN ERIC ERICHSEN. F. R. S., F. R. C. S., etc., Surgeon : —

" Experiments on living animals are absolutely necessary for the advancement of medical surgery and biological science. Such experiments should not be allowed without proper restrictions as a safeguard against their abuse by incompetent persons, or their being performed for futile purposes. Such experiments should only be performed for purposes of utility, — that is, the advancement of scientific knowledge, — and not for the purpose of acquiring manual dexterity ; nor should they be allowed as class demonstrations or for needless repetition. All experiments on living animals, if painful, should be performed under an anæsthetic.

" Experiments on living animals are most carefully restricted in this country. . . . I acted as government inspector of living animals for several years, and I can safely assert that the provisions of the act were vigorously enforced, and never, to my knowledge, contravened."

Prof. J. SULLY, London, England : —

" I believe both in the desirability of a certain control of vivisection experiments by the feeling of a community, or rather in a free expression of this feeling, and of urging it on legislators and others. At the same time, I think if an ignorant public were to determine when and where such experiments are to be carried out, some of the most valuable and useful parts of scientific work would be arrested. It is eminently a case for adjustment between a worthy popular sentiment and the claims of science. How these are both to be satisfied I am not clear, though firmly believing in the desirability."

From President FRANKLIN CARTER, LL.D., Williams College, Mass. :

"The second statement comes nearest to my opinion; but I have no belief in the desirableness of the State undertaking the delicate business of controlling or directing operations in laboratories. I do not sign the second statement, but send you this general reply."

From Prof. GEORGE A. COE, Ph.D., Professor of Philosophy, Northwestern University : —

(Professor Coe objects to signing any of the statements, and adds :) "The outcome of this consideration is not to let things take their course without reference to humanitarian considerations, or rather that humanitarian ethics should be persistently emphasized before all persons concerned, — students, instructors, faculties."

From JAMES R. CHADWICK, M. D., Boston, Mass. : —

" Vivisection is sometimes abused by the infliction of suffering upon the brute creation without corresponding benefit to humanity. These abuses I would like to see stopped. None of your plans seem to me to attain that object without putting objectionable restrictions upon this practice."

(Dr. Chadwick denies that it is a "common practice" in the United States to subject animals to torture merely for the purpose of demonstrating well known or accepted facts. The Committee beg to refer him to the earlier editions of Flint's Physiology, and to experiments upon the nervous system there detailed, as having been made "for class demonstrations.")

From VINCENT Y. BOWDITCH, M. D., Harvard Medical School, Boston, Mass. : —

" After weighing the matter carefully, I can. I think, conscientiously sign the second statement, ' Vivisection restricted by Utility.' I believe firmly that in the right hands vivisection has been and will be of great use to humanity. At the same time, I think that much (1 cannot say from experience how much) has been done that is not really necessary, and this I would have stopped."

From Rev. F. J. CLARE, Phillipsburg, Pa. : —

"It is questionable whether any prohibitory humane enactments can be fully enforced or will be fully obeyed. Nevertheless, right legislation is helpful and salutary, and should be encouraged."

T. M. BALLIET, M. D., Philadelphia, Penn. : —

"I have subscribed to vivisection restricted by utility, because I cannot see the advantage in vivisection as class demonstration; but fishing and gunning are recreations infinitely more cruel, and the pain inflicted on lower animals many hundred times larger, than the most unbounded vivisection by scientists."

From Prof. EARL BARNES, Stanford University, Cal. : —

"It is almost impossible to formulate a definition to which a thoughtful man can subscribe to-day. If I were to make my own statement, I should say that vivisection should be allowed where there is hope of discovery of some truth, or demonstration of some hypothesis of value to humanity, provided it is conducted by men who are wise and well trained."

From EPHRAIM CUTTER, M. D., LL.D., New York : —

" I am in favor of experiments made on criminals sentenced to death justly, for the good of society ; that is, in penal physiology. There is a crying need of the following experiments : (1) Feeding, to prove that tuberculosis is mainly caused by feeding, — and cured as caused, by feeding. (2) That fatty degeneration is caused and cured by feeding. (3) That thrombosis is caused and cured by feeding. (4) That cancer is caused and cured by feeding. (5) That all gravelly diseases come from, and can be made curable by, feeding. (6) That the noxious properties of common foods be tested on human criminals. (7) That consumption of the bowels be thus proved to come from food. (8) Finally, that worthless and criminal lives thus be made valuable for the good of society.

" I am utterly opposed to re-proving by vivisection facts already established. Many lives are now saved in gun-shot wounds of the abdomen because of vivisections on dogs, which otherwise would have been lost ; but crime should furnish the materials in most vivisections in the larger sense of the term."

From NATHAN E. BRILL, A. M., M. D., Visiting Physician of Mt. Sinai Hospital, New York : —

" I can only subscribe to statement III. in part, and to statement IV. (b) in part. I do not believe in the restriction to utility, nor in allowing unrestricted vivisection. We do not produce disease in man for the purpose of producing symptoms and pathological processes of that disease. Who would venture, for example, to inoculate a human being with small-pox for the purpose of showing its symptoms ? Students are ready enough to take the word of the teacher and their books as to the facts of that disease. Nothing of importance is gained by subjecting animals to torture to demonstrate established facts. Such experiments are wanton. . The truths of physiological laws should be taught, therefore, in the same way as are the facts of disease.

" Excepting these limitations, I am of the opinion that further restriction of vivisection would be a hindrance to scientific progress. . . . Vivisection should be taken away from those who use it simply to show generally accepted facts in connection with vital phenomena, because such experiments are useless. It seems to me to be an unjustifiable use of vivisection to apply it, as some physiologists do, simply to demonstrate facts which are universally accepted."

Sir DYCE DUCKWORTH, M. D., LL.D., Honorary Physician to H. R. H. the Prince of Wales : —

" A measure of State regulation is in my opinion desirable; but sentimentalists who have no means of forming sound opinions on the problems involved should not influence or bias the legislators. Vivisection should be forbidden to any but skilled teachers."

GEORGE FLEMING, C. B., LL.D., Ex-President of the Royal College of Veterinary Surgeons, North Devon, England : —

" I approve of this statement, ' vivisection restricted by utility.' "

From SAMUEL W. ABBOTT, M. D., M. A., Secretary State Board of Health, Boston, Mass. : —

"The definition of vivisection is a pretty difficult matter . . . I am in favor of its continuance, with such restrictions as shall make it impossible to conduct such operations in a cruel or painful manner."

From PAUL CARUS, Editor of "The Monist" and "The Open Court": —

"None of the four statements represents the opinion which I hold. I would formulate my position as favoring vivisection, with moral and without legal restrictions. While I am convinced that vivisection cannot be abolished, I insist upon its being restricted by the moral sentiment of the vivisector and his audience.

"The fundamental law of morality is not the avoidance of pain, but the intellectual and emotional growth and expansion of our souls. . . . Wherever you can detect professors and students who with blunted moral senses necessarily and cruelly inflict pain, you should at least call attention to the barbaric methods which they employ, even though they be scientists of high repute. To pass new laws will do no good; but to ventilate the question by public discussion, and help those vivisectionists who do not possess moral restraint to acquire it, will do an immense deal of good."

From MARY PUTNAM JACOBI, M. D., Visiting Physician at the New York Infirmary, Graduate from L'École de Médicine, Paris, France, 1871 : —

"The fourth statement entirely represents my views on this important question, yet leaves something to be added. It is in my opinion ridiculous for outsiders, necessarily imperfectly acquainted with methods of physiological research, to be allowed to prescribe what may or may not be done to demonstrate a proposition or to impart to students a living conception of the phenomena of life. But I think it quite fair that the experimenter should — like the butcher, and more than is at present the case with the hunter — be to a certain extent supervised and expected to reduce to a minimum the suffering inflicted, and to inflict none but what is absolutely necessary to attain his chosen end."

From MARY A. MIXER, M. D., formerly Professor of Physiology in Northwestern University, Woman's Medical College : —

"We consider any attempt to limit vivisection by the amount of pain produced to be entirely impracticable, since we have no means of knowing how much pain an animal suffers compared to a human being."

From Prof. ABBY LEACH, Vassar College, N. Y. : —

"Certainly, students in the laboratory ought to be held in check by their instructors, and it is unwarrantably cruel not to avoid pain where it is possible. If, however, the search for truth demands pain, *then have the pain.*"

From Rev. W. F. MUTCH, Ph.D., New Haven, Conn. : —

"With reference to the vivisection question, I am not disposed to sign any of the statements proposed. Doubtless, there are abuses in colleges as there are in barns; but it is clearly impossible to reach them with law, and it is one of the things which must be entrusted to humanity as it is."

Prof. BURT G. WILDER, M. D., Professor of Physiology, etc., Cornell University, Ithaca, N. Y. : —

" I cannot sign either of the declarations of your circular. My views and practice are indicated in the enclosed lecture-sheet."

· (Extract.) " From the use of a single word, ' vivisection,' for two widely different things, — painful and painless experimentation, — have resulted much confusion, injustice, and distress of mind. . . . Two kinds of vivisection should be verbally distinguished as sentisection and callisection. . . . Without prejudice to the claim of some that sentisection is demanded for the advancement of knowledge by experts, the writer holds that it is not warranted for the dissemination of knowledge."

MONCURE D. CONWAY, M. A., L. H. D., Author, England : —

" (1) I do not think that the question turns exclusively on pain ; even without pain, destruction of some animals should not be allowed without very great advantage, — for example, monkeys. (2) I do not think vivisection should be allowed for demonstration of facts already discovered. (3) It is doubtful whether, in philosophical research, the zoöphilists would be competent judges of the offices for which vivisection is justifiable."

From Rev. Dr. THOMAS C. HALL, Chicago, Ill. : —

" I believe the best thing the State could do would be to restrict the right of vivisection to responsible persons, and to control carefully the institutions under whose directions such persons worked."

From President G. STANLEY HALL, Clark University : —

" I hesitate somewhat between statements II. and III. I have no hesitation in dissenting from I. and IV. I do not quite wish to put myself on record as agreeing with either of the positions. It is very rarely justifiable to operate for mere demonstration, but I am not ready to see that absolutely forbidden on all animals. I believe most experiments can be made painless, but a few of the cardinal tests *cannot;* so that I should not have this absolutely forbidden.''

From H. H. A. BEACH, M. D., Surgeon, Massachusetts General Hospital : —

" I am unwilling to sign either of the four statements. Vivisection without restriction I regard as an abomination, and the laws which permit such a practice a blot upon the country that makes them. Vivisection restricted by utility I should sign, but there is no rule for excluding the clause, ' Science must prove that advantage and that necessity.' "

From Hon. J. S. WILLIS, M. C. : —

" It would be unwise and narrow to prohibit vivisection absolutely. It is allowable to mankind under restrictions to investigate all subjects that concern the welfare of the human family. . . . Utility should be a condition and prerogative of vivisection. The simple play of human fancy or curiosity should be discarded in testing and teaching so intricate a subject.

"There are other fields sufficiently wide and varied in which science may amuse itself, without attacking the citadel of animal life.

"It is doubtful if vivisection should be allowed at all, unless it can be performed without pain. Science is not warranted to inflict pain upon the basis of mere experiment. Where the use of anæsthetics can preclude the occurrence of pain, experiment for purposes of utility is allowable. Pain can never be banished from this world, and mankind must suffer its full share ; and it is unphilosophical to add to the horrors of commiseration by uselessly involving the inferior animals in the toils of exaggerated science."

From Pres. N. H. Chamberlin, A. M., LL.B., McKendrie College, Illinois : —

"I should be inclined to strike out in the second paragraph these words, 'but whether any useful knowledge can be acquired or not is beside the question.' I should rest the whole proposition on the moral quality of the act, without any statement of inducement implied in the clause struck out."

Sir WALTER BESANT, Author, England : —

"I am not a scientific man, nor can I test the statements and claims advocated in favor of vivisection ; but to cut and hack living creatures without the strongest possible reasons, or to suffer any irresponsible anatomist to do so without the strongest possible expectations, without anæsthetics, is unspeakably shocking and horrible."

RICHARD H. HUTTON, M. A., LL.D., Editor of the "Spectator," London, England : —

"Vivisection should be allowable if without any pain greater than is inflicted in putting to death in a humane fashion. This is the only answer which seems to me maintainable by any one who thinks it right to kill the lower animals for man's benefit ; and to this answer I adhere."

Hon. ARBA N. WATERMAN, Illinois Appellate Court, Chicago, Ill. :

"Civilization in its moral aspect consists in a heightened sympathy with and consideration for those men or animals in our power. It is impossible to train a child to indifference as regards the suffering of a helpless dog, and at the same time mindful of the rights of little children.

"It is immaterial whether he who proposes to torture be an ignorant savage or a distinguished savant. The aim of each is the same. The carter who pounds his horse, the boy who torments a kitten, and the scientist who twists the quivering nerves of a helpless dog are each, in his own way, endeavoring to promote 'human happiness.' Whoever believes his work to be of supreme importance will naturally become cruel. To whomsoever in the cause of science the agony of a dying rabbit is of no consequence, it is likely that the old or worthless man will soon be a thing which in the cause of learning may well be sacrificed. There is no reason for thinking that Torquemada or Robespierre were naturally any more cruel than the educators who endeavor to add brilliancy and piquancy to their lectures by an exhibition of the manner in which a dumb brute behaves when dissected alive. Vivisection should be permitted only to competent and trustworthy persons, and restricted to licensed places which shall be open at all times to inspection by the Presidents of Humane Societies for Protection of Animals, or their authorized representatives.''

OUIDA, Author, Florence, Italy : —

"Should ever such an opinion as that implied in the statement for vivisection without restriction become that of mankind in general, the world will be a hell indeed. The pretensions of what are called scientists are a menace to all liberty, peace, and virtue, and the doctrines thereof followed out from youth to age would make of the earth a shamble."

From H. R. BRISSETT, M. D., Lowell, Mass. : —

"Statements I. and II. I pass over as in no way tenable. Statement III. contains an " if." Anæsthesia is good while it lasts, but I have often seen it pass off, and the experimenter go on with the work in hand without renewing it ; and all the class saw with revulsion that there was real torture in the case. So I think absolutely, with such men as Tait and Bigelow, that vivisection subserves no good purpose, and has only theory (vague at that) to support it."

From CLAYTON L. HILL, M. D., Buffalo, N. Y. : —

" I know from personal experience that medical and surgical research does not demand the fearful suffering and waste of life that is entailed upon the lower animals. I have seen many hundreds of vivisections, and not one of them developed a new truth or an idea not already well established. Vivisection as practised in medical schools is simply a sort of theatrical performance for the benefit of the students and the glory of the professors."

From JOHN C. DU BOIS, A. M., M. D., Late Surgeon in the U. S. Army : —

" During three years of student life in Paris, I saw a good deal of torture to animals in unnecessary demonstrations to students of well-known facts, and I heard stories of Magendie's cruelties. A good deal of sentimental nonsense has been written and spoken upon vivisection ; yet there have been abuses, and the benefits such experiments have conferred do not palliate them. I am anxious for a proper and legal settlement of this subject."

From LUIGI GALVANI DOANE, M. D., New York City : —

" Put me down as the antagonist of vivisection in any form. The office of the physician is to heal wounds and to save life, not to take it."

From GEORGE H. PAYNE, M. D., Boston, Mass. : —

" I believe fully that we have no right to torture God's dumb creatures, and that it does little or no good to experiment upon animals."

J. W. THOMSON, M. D., New York City : —

" We do not believe that vivisection ever gave knowledge that led to the relief of a single human being from pain, or in any way helped to ameliorate human suffering. This diabolical practice is totally needless as well as dastardly inhuman. No man who has been guilty of vivisection ought to be allowed to practise as a physician. Imagine any one coming from a torture-chamber to see a sick child, or to have a mission to help suffering humanity! How can one who is callous to animal suffering yearn to help his fellow-man? What can be learned from the quivering, writhing flesh for intelligent guidance for the sick? Do the tissues, laid open by the lance, display normal function? Assuredly not. What is the story that a humane mind would read? The only true one of which we can conceive is the palpitating plea for mercy, unheeded by the inhuman wretch who, in the name of a false science, gloats like a ghoul over his fiendish, bloody work."

From JAMES P. HAWES, M. D., North Hector, N. Y. : —

" . . . I have seen dogs kept howling and starving, with large doses of chloral injected in one side and strychnine in the other side; some lived and some died. What did it prove? In one of our leading colleges, I have seen a calf split open and a little flag-staff stuck in the heart for the edification of the students! What did that prove of the human heart? Enough accidents happen to poor humanity to test the results of pain and wounds, burns and scalds, freezing and anæsthesia, without infliction on the poor brute."

From A. ROSE, M. D., New York: —

" There can be no nobler cause than the prevention of cruelty to animals in vivisection. During the Middle Ages tortures were inflicted under the very eyes and strict supervision of awfully learned physicians, and thus we see that learning does not prevent us from doing cruelty. We have to be reminded of this example."

CHARLES W. SUPER, President of Ohio University : —

" While I am not quite ready to say that vivisection should be entirely prohibited, I am very strongly inclined to this opinion ; but I am entirely ready to say that scientific research ought not to be free and untrammelled. The humanitarian interests of the world are paramount to any and every other. Scientific investigators sometimes become veritable monomaniacs in their search for knowledge, and as indifferent to the highest interests of their fellow-men as other lunatics. If under the ' untrammelled-research ' régime, my neighbor seems to be a better subject for investigation than his dog, I ought as a matter of consistency to experiment on the man rather than the brute. It is not hard to see whither such consistency would lead. I am glad to see that the vigorous agitation of this subject is leading to what I consider wholesome legislation."

Hon. ROWLAND B. MAHANY, M. C. : —

" Enough cruelty is already being practised upon dumb animals without legalizing what at best may be termed ' experimental torture.' . . . Even under the scalpel of the ablest operator, it is a grave question of doubt whether any permanent benefit to science is acquired by the process of vivisection."

Right Rev. GEORGE F. SEYMOUR, LL.D., Bishop of Springfield :

"I consider that the animated world beneath man is a sacred trust committed to him by the Creator, and for the right and just administration of which he will be held sternly accountable. Hence, I would place very severe restrictions upon vivisection, and allow its practice only in cases where it was employed for settling questions which we have good reason to believe could not be answered except by such experiments. I would exclude absolutely, and forbid under penalty by law, all exhibitions to students of vivisection as illustrating ascertained and recorded facts of science."

From President W. P. JOHNSTON, Geneva College, Pa. : —

"My opposition to vivisection is not so much because of the pain to the animal dissected (it dies in a little while), but because of injury to the moral nature of the animal dissecting, that lives probably for many years, and has *other* chances on other animals than dogs and cats !"

From John H. KEYSER, HARTFORD, Conn. : —

"I was superintendent of the Stranger's Hospital in New York at a time when vivisection was freely practised upon animals by young student physicians. From that experience I formed the opinion that it was wicked, wanton, and cruel to clothe these young and inexperienced men with despotic power over animals, and I forbade the practice. Mercy towards the helpless brute creation is, in my judgment, ample argument against vivisection, and it ought to cease."

From John BOARDMAN, M. D., Buffalo, N. Y. : —

"I do not believe that any real good comes to the ordinary student from class vivisection. On the other hand, in my opinion, the humane part of man is injured and the diabolical part comes nearer to the surface."

From THOMAS B. FOWLER, M. D., Cohocton, N. Y. : —

"I do not think the impression left on the mind of the average medical student as a result of witnessing the mutilation of animals is one that really adds to his available store of knowledge, or tends to aid him in prescribing for suffering humanity. It is impossible to estimate the baneful effects of such experiences on the minds of men whose actions are largely governed by impulse or propensity."

NOTE. — The Committee greatly regret that considerations of space prevent further quotations from the numerous letters received. It is possible that many of them may be hereafter printed in another form.

www.ingramcontent.com/pod-product-compliance
Lightning Source LLC
Chambersburg PA
CBHW022001190326
41519CB00010B/1355